RUNNING SMALL MOTORS WITH PIC® MICROCONTROLLERS

About the Author

Harprit Singh Sandhu, BSME, MSCerE, is the founder of Rhino Robotics, a manufacturer of educational robots, computer numeric controlled machines and the software to control them. Rhino provided the first truly integrated vision system for robots as a part of the RoboTalk robot control language. He is the author of *Making PIC Instruments and Controllers* (McGraw-Hill/Professional, 2008).

RUNNING SMALL MOTORS WITH PIC® MICROCONTROLLERS

Harprit Singh Sandhu

New York Chicago San Francisco Lisbon London Madrid
Mexico City Milan New Delhi San Juan Seoul
Singapore Sydney Toronto

The McGraw·Hill Companies

Library of Congress Cataloging-in-Publication Data

Sandhu, Harprit.
Running small motors with PIC microcontrollers / Harprit Singh Sandhu.
p. cm.
ISBN 978-0-07-163351-2 (alk. paper)
1. Microcontrollers. 2. Electric motors—Electronic control. 3. Programmable controllers. I. Title. II. Title: Running small motors with Peripheral Interface Controller microcontrollers.
TK2851.S26 2009
621.46—dc22 2009023986

McGraw-Hill books are available at special quantity discounts to use as premiums and sales promotions, or for use in corporate training programs. To contact a representative, please e-mail us at bulksales@mcgraw-hill.com.

Running Small Motors with PIC® Microcontrollers

1 2 3 4 5 6 7 8 9 0 DOC DOC 0 1 9

ISBN 978-0-07-163351-2
MHID 0-07-163351-0

Sponsoring Editor
Roger Stewart

Editorial Supervisor
Patty Mon

Project Manager
Smita Rajan,
International Typesetting and Composition

Acquisitions Coordinator
Joya Anthony

Copy Editor
Sally Engelfried

Proofreader
Mohinder Bhatnagar,
International Typesetting and Composition

Indexer
Robert Swanson

Production Supervisor
George Anderson

Composition
International Typesetting and Composition

Illustration
International Typesetting and Composition

Art Director, Cover
Jeff Weeks

Cover Designer
Jeff Weeks

This effort is dedicated to
Martin Donald Ignazito
Engineer and Gentleman.
I've known Marty almost as long as I've known anyone. We were in school together at the University of Illinois at Urbana, IL, and he was my partner when I was in the engineering business. We have been friends for well over 45 years. He is one of the best engineers I have ever come across and can provide a well thought out approach to almost any engineering problem in short order. Since he retired he has become an avid para-wing aviation enthusiast and an expert on the selection of propellers. He also provides instruction in these machines. He has helped author some of the FAA standards for light aircraft. We have spent many, many good times together.

CONTENTS AT A GLANCE

CONTENTS

PREFACE

How It Happened

In 1995, I put in the public domain an outline of what would have to be added to the Meccano and Erector set systems to allow aspiring young engineers to make sophisticated automatic machines of all kinds with these systems. These systems provide almost everything imaginable in the way of mechanical components and just about nothing in the way of electronics. Adding electronics would change everything. Everything! I imagined the electronic engineers would take it from there and soon there would be a comprehensive electric/electronic system we could all use. To say I was wrong would be more than an understatement.

I had lots of correspondence from enthusiasts all over the world telling me what a great thing this would be, but no one seemed interested in providing what was needed. If this was going to happen, it was up to me and I was going to have to learn how to do it! Since I was then employed full time, I did not have the time to create this system. However, I am now retired and have taught myself what is needed to run motors with microcontrollers. In this book I share what I have learned with you. I will be putting motor amplifiers and other components on the market as I develop them. My initial work in this direction is described herein.

If you want to take a look at what I have to say about the standard I described, it is on the Internet at www.pinecreekbay.com/harpritsan/MeccanICindex.html.

This tutorial introduces you to the basic techniques used to run small DC, DC servo, stepper, and R/C servo motors with microcontrollers. It concentrates on using the microcontrollers made by the Microchip Corporation, with particular emphasis on the 16F877A and 18F4331 40-pin microcontrollers. It uses microEngineering Labs' LAB-X1 board to make things easier for the experimenter, but you do not need to have the board to learn to do what needs to be done. Other MCUs and microprocessors made by other manufacturers can also be used. They are similar, and techniques similar to those developed herein are used. (Running larger motors is essentially a matter of using more powerful amplifiers; the techniques described herein for running them are the same.)

Motor control can take on a number of forms from simple on/off control to carefully managed intricate motion profiles. The language used to control the motor can vary from assembly language and C to PICBASIC PRO. We will go through all the techniques that are suitable for this introductory text with the PICBASIC PRO language. Beginners will find that routines written in PICBASIC are much easier to understand than those written in other more primitive languages. Once you understand the basic routines, they can be written in the language of your choice. Conversion of the routines developed herein to assembly language or the C language can be undertaken by those interested in doing so with relative ease but will not be undertaken in this text.

PIC Microcontrollers

I selected the Microchip Technologies family of PIC microprocessors as the focus of these notes for two reasons. First, the Microchip provides the most comprehensive line of microprocessors for the kind of projects we are interested in. Second, the compiler for these processors provides almost the entire line of PICs with comprehensive support. All you have to do is tell the compiler which PIC you are using, and if the features you have been addressing in your project are available on that PIC, the compiler will do the rest. You will never have to buy another compiler if you stay with the very comprehensive Microchip Technologies family of PICs.

Other Microprocessors

We will be using a PIC 16F877A and 18F4331 for all experiments, but any number of microcontrollers are suitable for the task. The selection that you make will depend on the availability of suitable software support and other features that you need on the MCU for your particular application.

THE 16F877A AND THE 18F4331

The first part of the book concentrates on giving you a basic understanding of how a typical microcontroller works, with focus on the PIC 16F877A. Once you know how the 16F877A works you will be able to use other similar microprocessors with relative ease.

Enough is covered about the 18F4331 to allow you to use its ability to keep track of what is going on with the standard quadrature encoder interface attached to the motor. This PIC was selected primarily so we can use its ability to keep track of the encoder counts autonomously.

The second part of the book covers the use of the microcontrollers to run the small motors that we are interested in. (Larger motors need larger motor power amplifiers, but the control techniques are similar.) The following motors are covered:

- Model aircraft R/C servos
- Small, plain DC motors
- Servo DC motors with encoders attached
- Stepper motors (bipolar)
- Small AC motors and solenoids

All the material is covered in a nonmathematical way so that anyone interested in learning to run motors can learn to do so with a minimal technical background.

I could not have created this book without the patient help of Charles Leo at micro-Engineering Labs, Inc. Though I never met him, Charles answered countless e-mails from me without protest and with extreme patience (as I discovered, some of my questions were not the most enlightened).

Should you discover errors in the tutorial, I would appreciate receiving an e-mail description of the error so that I can make the necessary corrections.

I handle all customer support personally, and you are welcome to e-mail me with relevant questions, comments, and corrections. You can contact me at harprit.sandhu@gmail.com.

—Harprit Singh Sandhu
Champaign, Illinois

Internet support sites: Encodergeek.com and www.mhprofessional.com/sandhu

Part I

MICROCONTROLLERS

We need to understand what one specific microcontroller can do in some detail so we can use it effectively to control motors.

1

INTRODUCTION TO MICROENGINEERING LABS' LAB-X1 EXPERIMENTAL BOARD

A vast array of PIC (Peripheral Interface Controller) microcontrollers is manufactured by the Microchip Technology Corporation of Tucson, Arizona. Microchip has shipped over ten billion of their devices all over the world. They are everywhere. Learning to use them is both easy and enjoyable and will serve you well if you are a student, a hobbyist, or an engineer or if your work involves the use of microcontroller-based devices.

This tutorial is designed to introduce you to these devices as they apply to running motors. I intend to do this in a nonintimidating way for the technically inclined who are not necessarily electronic technicians or electrical engineers.

We need to have a comprehensive understanding of and familiarity with at least one microcontroller in the rather large family of PIC microcontrollers if we are going to use them for the sophisticated control of all sorts of motors. I picked the PIC 16F877A because it provides almost all of the many features found in microcontrollers that are made by the many suppliers of these small yet comprehensive logic engines.

As novices, if we want to get familiar with running motors with microcontrollers, we need an easy to use yet sophisticated and versatile board to play with and test our ideas on. Though of course it is possible to design and build a board that would do that, we do not have the expertise to do that at this time. I selected the very popular LAB-X1 and the related PICBASIC PRO compiler software as the basic platforms for the projects and ideas presented in this book. As you go through the book, you will find that the system provides an easy to use and versatile platform for checking out your hardware and software ideas before committing to printed circuits, wire, and solder. microEngineering Labs, Inc., the manufacturers of the LAB-X1 board, maintain a very useful and helpful web site (www.microchip.com) that will be a tremendous aid for you as you learn about your LAB-X1 in particular and the Microchip

Technology Corporation PIC microcontrollers in general. Their web site contains a large number of example programs, tutorials, and other technical information that will help you get started with using these microprocessors. There are also a large number of other web sites that are dedicated to the support of PIC microcontrollers.

This book supplements the information on the Internet from the microEngineering Labs site and from other sources. We will use the sample programs (modified for clarification as may be necessary) and other information that is on the web. The book provides extensive diagrams in a format that you can use to help you design your own devices, with minor modifications, based on what you learn.

There are two basic aspects of PIC microcontrollers: hardware and software. The LAB-X1 board is designed to provide you with the hardware platform you need to conduct your first software and hardware experiments with PIC microcontrollers, specifically the 40-pin family subset. The PICBASIC PRO (PBP) compiler, provided by the manufacturers of the board, programs this and similar microprocessors and is easy to use and powerful; the code created is fast and efficient.

If you have a serious budgetary constraint, the software for use with this board is the Basic Compiler from microEngineering Labs. This compiler is available for about $100 (in 2009), but I don't recommend it for serious work.

On the other hand, if you have a serious interest in using PIC microcontrollers, especially if you will be using them for a long time, I recommend the PICBASIC PRO compiler because it gives you the comprehensive power and ease of use that you need to rapidly perform useful everyday work. The PRO compiler is available for about $250 (in 2009), and all the software discussed in this workbook was written for the PICBASIC compiler. A listing of instructions and keywords provided with each compiler is provided in Chapter 4.

You can get a free, limited copy of the PBP (picbasic pro) compiler on the Internet on the microEngineering Labs web site. This copy contains all the instructions in the full version of PBP but is limited to 30 lines of code. Even so, it can be used to effectively try out the powerful command structure of the language. The instructions for the language can also be downloaded from the microEngineering Labs web site at no charge. Before you make a decision about your compiler purchase, try out the free version.

You will also need a hardware programmer to allow you to transfer the programs you write on your personal computer (PC) to your PIC microcontroller. Programmers are also available from microEngineering Labs for the parallel port, the RS232 serial port, and the USB port of your computer. These programmers make it a one-button click to transfer your program from your computer to the microcontroller and to run it without ever having to remove the MCU (micro controller unit) from the board. I recommend the USB programmer.

The software needed to write and edit the programs before transferring them to the programmer and onto the microcontroller is a part of the compiler package. Other editors are available at no charge from a number of other suppliers. Programs can also be written in Microsoft Word and then cut and pasted into the programming software.

The salient hardware features (with some repetition by categories listed) provided on the LAB-X1 are listed next. The following input capabilities are provided:

- A 16-switch keypad, plus a Reset switch
- Three potentiometers
- IR (infrared) detection capability, no detector provided
- Temperature sensing socket, no IC (integrated circuit) provided
- Real time clock socket, no IC provided
- Sockets for experimenting with three basic styles of one-wire memory chips
- Serial interface for RS232, IC provided
- Serial interface for RS485, no IC provided
- PC board holes are provided for other functions. See the microEngineering Labs web site for further details.

The following output capabilities are provided:

- Ten LED bar graph with eight programmable LEDs
- 2-line x 20-character LCD display module
- A piezo speaker/beeper
- DTMF (dual-tone multi-frequency) capability (digital tones used by the phone company)
- PWM (pulse width modulation) for various experiments
- IR (infrared) transmission capability, no LED provided
- Two hobby radio control servo connectors, no servos provided
- As mentioned, sockets for experimenting with:
  - Serial memories
  - A to D conversion with 12-bit resolution
  - Real time clocks

The following I/O interfaces are provided:

- RS232 interface
- RS485 interface, socket only (the interface chip is inexpensive and easy to obtain)

You can investigate the use of the following three types of serial EEPROMs:

- I2C
- SPI
- Microwire

The following miscellaneous devices are also provided:

- Reset button
- 5-volt regulator

- 40-pin ZIF (zero insertion force) socket for PIC micro MCU (the recommended PIC 16F877A IC is not provided)
- Jumper selectable oscillator from 4 MHz to 20 MHz
- In-circuit programming/debug connector
- Prototyping area for additional circuits
- 16-switch keypad
- Socket for RS485 interface (device not included)
- Socket for I2C serial EEPROM (device not included)
- Socket for SPI serial EEPROM (device not included)
- Socket for Microwire serial EEPROM (device not included)
- Socket for real time clock/serial analog to digital converter (devices not included)
- Socket for Dallas 1620/1820 time and temperature ICs (devices not included)
- EPIC (Epic is a trade mark of microEngineering Labs, they give no explanation) in-circuit programming connector for serial, USB or parallel programmer
- A small prototyping area for additional circuits

All in all, it's a very comprehensive, well thought out, and useful experimental platform. The board is available assembled, as a kit, or as a bare PCB; see Figure 3.1. The board is 5.5" x 5.6".

As already mentioned, not all the features I mentioned here are completely implemented, but sockets or PC board pin holes are provided for all of them. You may not have to make any soldering additions to the board to use the features you are interested in, but you do have to purchase the additional IC chips if you want to use them. The standard version of the board as shipped to you includes the following:

- The assembled board
- Software diskette, which includes:
  - PDF schematic of LAB-X1
  - Sample programs
  - Editor software

The 40-pin PIC microcontroller is not included. As received, the board is configured to run a 4 MHz, but it can go up to 20 MHz.

THE MICROCONTROLLER

The PIC 16F877A microcontroller (which is a necessary component on the board) is not provided because each of the compatible PIC microprocessors available has varying features, and you need to select a unit that suits the application that you have in mind. We will be using the recommended PIC 16F877A and 18F4331 microcontrollers for all our experiments. If you want to use a different processor, be sure to check for pin-to-pin compatibility on the web. Data sheets can be downloaded for all the

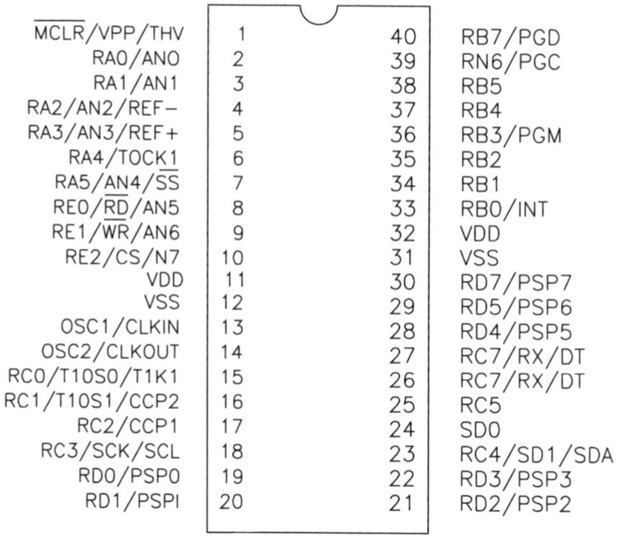

Pin name	Pin	Pin	Pin name
MCLR/VPP/THV	1	40	RB7/PGD
RA0/AN0	2	39	RN6/PGC
RA1/AN1	3	38	RB5
RA2/AN2/REF−	4	37	RB4
RA3/AN3/REF+	5	36	RB3/PGM
RA4/TOCK1	6	35	RB2
RA5/AN4/SS	7	34	RB1
RE0/RD/AN5	8	33	RB0/INT
RE1/WR/AN6	9	32	VDD
RE2/CS/N7	10	31	VSS
VDD	11	30	RD7/PSP7
VSS	12	29	RD5/PSP6
OSC1/CLKIN	13	28	RD4/PSP5
OSC2/CLKOUT	14	27	RC7/RX/DT
RC0/T10S0/T1K1	15	26	RC7/RX/DT
RC1/T10S1/CCP2	16	25	RC5
RC2/CCP1	17	24	SDO
RC3/SCK/SCL	18	23	RC4/SD1/SDA
RD0/PSP0	19	22	RD3/PSP3
RD1/PSPI	20	21	RD2/PSP2

Figure 1.1 The 40-pin 16F877A PIC microcontroller.

microcontrollers at no charge from the Internet. The commonly used 40-pin for pin-compatible MCUs are the 16F873, 16F874, 16F876, 16F877, 18F4331, and 18F4431. They share similar power and pinout layouts but exhibit different capabilities. Other PICs may also be used.

The following 40-pin PICs will work in the LAB-X1

PIC 16C64(A), 16C65(B), 16C662, 16C67, 16C74(AB), 16C765,
16C77, 16C774, 16F74, 16F747, 16F77, 16F777,
16F871, 16F874, 16F874A, 16F877, 16F877A, 16F914,
16F917, 18C442, 18C452, 18F4220, 18F4320, 18F4331,
18F4410, 18F442, 18F4420, 18F4431, 18F4439, 18F4455,
18F448, 18F4480, 18F4510, 18F4515, 18F452, 18F4520,
18F4525, 18F4539, 18F4550, 18F458, 18F4580, 18F4585,
18F4610, 18F4620, 18F4680

SOFTWARE COMPILER

The PICBASIC PRO BASIC software compiler provided by microEngineering Labs provides the functions needed to control all aspects of the hardware provided by Microchip Technologies as a part of their large PIC offering. All the functions available on the PIC 16F877A microcontroller that we will be using are accessible from the software. The PICBASIC software will write software for almost the entire family of PIC microcontrollers. You will be able to use this compiler for all your future projects; it is a very worthwhile investment.

ADDITIONAL HARDWARE

The following hardware can be added to the LAB-X1 without making any modifications to the board. These hardware items fit into sockets or onto pins that are provided on the LAB-X1 as shipped. Not all devices can be mounted simultaneously because some addresses are shared by the sockets provided. In our experiments, we will populate only one of the empty sockets at a time, to make sure that there are no conflicts. (There is no need to use more than one device at one time for any one experiment so this will not be a problem.)

Memory chips:

- I2C memory chip
- SPI memory chip

Microwire memory chips:

- 12 bit A to D converter chip
- NJU6355

Real time clock chips:

- DS1202
- DS1302
- LTC1298

Thermometer chip:

- DS1802

Serial interface chip:

- RS485

RC servos:

- Two hobby R/C servos can be controlled simultaneously; not provided.

The LAB-X1 provides two sets of pins for the R/C servos. All standard model aircraft servos can be used and you can use either one or two of them. (Using these is essentially an exercise in creating pulse width modulated signals and profiles that are used in the R/C industry.)

40-PIN DEVICES

All 40-pin MCUs provided by Microchip can be accommodated in the 40-pin ZIF socket provided on the board. Check for compatibility with the pin layout before selecting and buying your MCU. The recommended PIC 16F877A that we are using is an excellent choice for learning if you have no specific use in mind.

We will also be using the 18F4331 for the experiments needing encoder interfacing with the microprocessor. This chip has the ability to keep track of the encoder position automatically, which is a very useful property for our purposes.

BREADBOARDING AND EXPANSION

All 40 pins of the MCU have been provided with extra predrilled PC board holes. These can be used to extend the signals from these pins to an off board location for experimentation. The extensions are easily made with standard 0.1 inch on center pins and matching cables and headers.

A small breadboard space is provided on the LAB-X1 itself to allow the addition of a limited number of hardware items that you may need to experiment with.

See the Internet support web site www.encodergeek.com for availability of ready-made headers and cables and so on for use with the LAB-X1.

SPECIAL PRECAUTIONS AND NOTES OF INTEREST

The following caveat could have been placed later in the book but is included here to encourage you to select the programmer best suited to your needs.

Pin B7 on the LAB-X1 is connected to a programming pin on the EPIC parallel programmer at all times, and the programmer forces this pin high. If you are using this pin in your experiment and you need to have it be low, you must disconnect the EPIC programmer to release the pin. The major benefit of using the parallel programmer is that it frees up your computer's serial port for communications with the LAB-X1, but if you are using a USB programmer, it can be left connected to the LAB-X1 at all times. This is the reason I recommend the USB programmer.

Resistor R17, which is connected to the keypad, is of no consequence to the operation of the LAB-X1. It is needed for some PIC programming functions and can be ignored for our purposes.

DATA SHEETS

The hardest part of using these microcontrollers is understanding the huge data sheets—often 400 pages or so. Since each data sheet is similar but different from every other data sheet, you are advised to select one or two microcontrollers to get familiar with and use them for all your initial projects. In this workbook the three that are discussed are the PIC 16F84A (this chip will not fit in the 40-pin socket provided but is a good alternate choice) for your small projects and the PIC 16F877A for larger, more comprehensive projects. Each of these uses flash memory and can therefore be programmed over and over again with your programmer and a programming socket. The processor you select will be determined by the kind of I/O and internal features that you need and the availability of inexpensive OTP (one-time programmable) equivalents if you are going to go into production. We will use the 18F4331 also but only for the encoded motor experiments.

A lot of the information in the data sheets is more complicated and detailed than we need to worry about; we can do a lot of useful work without understanding it in every detail. For example, the timing diagrams and other data about the internal workings of the chips are beyond what we need to understand at the level of this book. Our main interest should be in what the various registers are used for and how to use them properly and effectively, as well as being able to set the various registers in the system so that we can activate the features we need for each particular project. Understanding timers and counters is a part of this. The entire interaction of the microcontroller with its environment is determined by the I/O pins and how they are configured, so knowing how to configure the I/O competently is very important.

The data sheets are available as PDF (portable document format) files on the Internet from the microEngineering Labs web site or from the Microchip web site. Download these onto your computer for immediate access when you need them. Keeping a window open specifically for this data is very handy, but you will also want to print out some of the information to have it in your hands.

The areas of the data sheet that support our needs are the following:

1. Understanding and becoming familiar with what has already been defined by the compiler software as it relates to the software
2. Getting familiar with the addressing and naming conventions used in the data sheet
3. Understanding the use of the various areas of memory on the MCU
4. Learning how to assign and use the I/O pins to your best advantage
5. Understanding how to use the PBP software to its best advantage and writing programs that are as fast as possible
6. Getting familiar with the register naming conventions and usage.

ANOTHER INTERESTING BOOK

David Benson of Square 1 Electronics wrote a very interesting and useful book on the PIC 16F84A called *Easy Microcontrol'n* (this book used to be called *Easy PIC'n*) that supports these investigations. It taught me a lot of things I did not know and had not even thought about. In this workbook you will reap some of the benefits of my learning experience. I recommend that you get a copy of *Easy Microcontrol'n* to support your use of the PIC 16F84A. It has a lot of very useful information in it and will save you a lot of time and headaches. However, the book is comparable to a first course at the community college level, and I found it too dry, with the emphasis on doing things without a BASIC compiler. A BASIC compiler is the easy to use tool of choice in this workbook because of our interest in getting things done in a hurry as opposed to becoming PIC/MCU experts in assembly language. The emphasis here is more in applied results rather than rigorous foundation level learning of assembly language programming. This does not in any way negate the usefulness of Benson's book to those interested in understanding and using the PIC 16F84A and similar microcontrollers.

Caution *All the programs in Benson's book are in Assembly Language.*

A FAST INTERNET CONNECTION IS A MUST

You absolutely must have an Internet connection because so much of the information you need is on the Internet. It is very helpful to have more than a standard phone line connection so get the fastest connection you can afford. A cable modem is strongly recommended. If you and a couple of neighbors can get together and form a local area network (LAN) and share a wireless (Wi-Fi) modem setup, it becomes a really inexpensive way to get fast Internet service. The Wi-Fi signals have no problem reaching all the apartments in a small building and sometimes even the house next door. Amplifiers and repeaters are available to increase signal strength where necessary.

DOWNLOADING DATA SHEETS

One of the first things you need to download is the data sheets for the PIC 16F87X. You will in all probability end up using the smaller and less expensive PIC 16F84A for a lot of your initial projects, so it might be a good idea to download the information for that microcontroller while you are at it. As mentioned before, these files are available from the Microchip web site and the information is free. However, the two documents are about 400 pages all together, so you probably will not want to print it all out. You will, however, want to print some of the more commonly used information so you can refer to it whenever necessary. The rest should be stored on your computer so that you can call it up or search for what you need when you need it.

The Microchip Technology Corporation website is at www.microchip.com.

Finding what you need will be under "support" on their web site, and it is easy to download. Just follow the instructions provided on the site.

2

GETTING STARTED

The Hardware and Software

This chapter lets you know what you need in the way of minimum hardware and software to get started and what you need to do to set it up and get it ready for use.
List of hardware and what comes with it

- The LAB-X1 board (with software CD ROM)
- Power supply for LAB-X1 (with wall-mounted transformer)
- Serial port, parallel port, or USB port programmer for the board (with software CD ROM connection cable for LAB-X1 board [10-pin], cable to go from computer to programmer)
- Power supply for the programmer (with wall-mounted transformer)
- PIC 16F877A microcontroller or equivalent (see list in Chapter 1)

List of the required software

- PICBASIC PRO compiler
- MicroCode Studio editor software for writing the programs

List of the required information

- Data sheet for PIC 16F877A microcontroller downloaded from the Internet or the CD

List of computer equipment you already should have

- 1Wintel computer (IBM-PC or compatible with hard drive), CD reader (needed only to read software on CD ROMs but nothing else), printer, Windows operating system, and access to the Internet. (A broadband connection is strongly recommended.)

The Programmers

microEngineering Labs offers three programmers. One uses the parallel port, one uses the USB port, and the third uses the serial port. The operation of the three programmers is almost identical as far as the user interface is concerned. In this book we will use a USB programmer for all our experiments; this is what I used. The new USB programmer is more convenient to use than the other programmers because it does not need a power supply; it gets its power from the USB port. An important bonus is that it frees up the COM port for use with the computer (the parallel programmer does this also).

BUILDING A PROGRAMMER

There are a number of plans on how to make inexpensive programmers on the internet, but I am not going to recommend any of them because I have not built any of them.

USING THE PROGRAMMERS

The USB programmer does not need a power supply or wall transformer. It gets its power from the USB port. Using a USB port frees up the serial port for your experimentation and this is important because most of the new computers have only one serial port. The PC serial port connects to the LAB-X1 serial port for certain uses.

For the serial port and parallel port programmers, first plug the 16-V power cord connector into the programmer and then into the wall socket. The USB programmer needs to be connected but does not need a power supply connection. If you do not have power to the programmer when you start the programming software, the software will not be able to see the programmer and an error message will be displayed: the software will report that it could not find the programmer.

It is best to start the programmer software from the MicroCode Studio Editor window. If you do it this way the microcontroller being used is selected automatically and the program you are working on in the MicroCode Editor window is transferred automatically to the compiler software and onto the MCU on the LAB-X1 board. It can all be set up to be a one-click operation. See Appendix A.

If you are programming an MCU that is not on the LAB-X1, insert the microcontroller into the programming socket immediately before you begin programming the microcontroller. This applies only if you are programming a loose microcontroller. If you are programming a microcontroller plugged into the LAB-X1, it can be left in the board all the time.

Caution *The only exception for the parallel port programmer is that the B7 pin is pulled low by this programmer and will interfere with your program if you are using the B7 pin. If you are going to be using this pin, you must unplug the programmer between programming sessions.*

The sequence to create a program inside a microcontroller is as follows:

1. Write program in the MicroCode Studio Editor environment.
2. Compile the program.
3. Program the device.
4. Use the device.

The last three steps can be combined into one keystroke. See Appendix A.

Loading the Software

The following software will be provided with the various components that you will acquire as you learn about using microcontrollers based on the experimental boards provided by microEngineering Labs.

- PICBASIC PRO compiler software and book.
- USB port programmer (or whatever programmer you are using) software and book
- MicroCode Studio (the editor) on CD ROM or downloaded from the Internet.

The DOS environment is archaic and can be difficult for users not familiar with it. You do not have to deal with DOS to use and enjoy the hardware and software that we will be using. Everything can be done from the Windows environment.

Note *If you need to use DOS, there is a chapter at the beginning of the manual that tells you what you will need to do.*

The PICBASIC PRO compiler manual covers the use of the software in the DOS environment. I suggest that you ignore the first pages of the book and instead read the following section on how to run everything under the Windows environment. Once you are familiar with how the system works, you can go back and learn how to use the software in the DOS environment. There are a number of things that the DOS environment provides that can be useful and you will want to know about these as you get more and more proficient in your use of the microcontrollers.

USING THE SOFTWARE IN THE WINDOWS ENVIRONMENT

The first question that needs to be answered in almost every endeavor is always: "What do I need, what do I have to do, and what will it cost me to get the job done?" Accordingly, we will address this now.

Let's assume that you already have an IBM-PC with a suitable Windows operating system and that you know how to use it. Your computer needs the following capabilities to allow you to access the hardware and software that you are going to use it with.

In this book I will deal exclusively with the IBM-PC in a Windows environment. The software is not available for the Macintosh. Here is what you need:

- A 3.5 inch floppy drive or CD ROM (as of this writing some of the software is provided on 1.4 Mb, 3.5 inch floppies only and you must read it off a diskette for the system to work right. You cannot copy the software to a CD ROM and work from there; it will not work.) If your software comes on CDs, you can ignore this.
- A hard disk with about 5 MB of free space for software storage and as a general workspace.
- A serial port (COM1 or COM2) if you will be using the new serial programmer, and a USB port if you will be using the USB programmer. The USB programmer has the advantage of not needing a wall transformer based power supply because it takes its power from the USB port.
- LAB-X1 Experimenters board ($195); 16F877A microcontroller (not part of Lab-X1) ($10).
- USB programmer ($120).
- PICBASIC PRO compiler ($250).
- Miscellaneous motors and electronic items for experimentation ($70) allowance.

The allowance for the motors and electronic parts also covers the need to purchase memory- and time-based components that are socketed for but are not listed. You may decide that you do not need to experiment with some of these at this time. The allowance provides for almost everything you need for your motor experiments.

microEngineering Labs also provides a number of other preassembled boards for experimentation and educational purposes that you should be aware of:

- Lab-X1 Experimenter's board, which we are discussing
- Lab-X2 Experimenter's board for custom circuits
- Lab-X20 Experimenter's board for 20-pin devices
- Lab-3 Experimenter's board for 18-pin devices
- Lab-4 Experimenter's board for 8 and 14-pin devices
- Lab-XT Experimenter's board for telephone technology–related investigations
- Lab-XUSB Experimenter's board for building USB interfaces and peripherals

In this book we will consider the LAB-X1 only. This board provides a 2-line by 20-character display, which is very useful in the learning environment because it can allow you to see what is going on in the system as you experiment (if you program your programs to do so). Since almost all the microEngineering Labs boards provide similar features, learning transfer to the other boards is high.

Start out by opening a new folder on your desktop and labeling it LAB-X1 Tools. You will store everything that has to do with all your projects in this folder. You are opening this folder on the desktop now, but you can move it to wherever you like in the future. For now, you don't have to make a decision about where to locate the folder, and it is right in front of you when you start your computer and the desktop appears.

Open the LAB-X1 Tools folder and create new folders, one for each of the items or applications we will be working with in this folder. Name these folders as follows:

- MicroCode Studio
- USB Programmer (or whichever programmer you decide on)
- PICBASIC PRO compiler
- LAB-X1 and related information

Then follow these steps:

1. Put the MicroCode Studio CD in the disk drive and open it.
2. Copy all files to the MicroCode Studio folder.
3. Eject the CD and put it away in a safe place.
4. Put the programmer diskette in the disk drive and repeat the steps that were taken above for the software in this package. Repeat the process for all the diskettes.
5. Put a shortcut for the MicroCode Studio program on your desktop. This is the only shortcut you need when you want to create programs for your MCUs. All other functions of the system can be accessed from the window of this editor.

As a general rule, you will never see the compiler as such. It is called from the MicroCode Studio Editor screen, it works on compiling the program it is asked to compile, and then it disappears into the background ready for the next compilation request. The errors that are displayed after a compilation are generated by the compiler. If all goes well, there are no errors and you get a message telling you that the compilation was successfully performed. The new hex file just generated will appear in the directory listing the next time you open a file. The hex file will have the same name as the text file that it was compiled from. The PICBASIC PRO compiler manual covers how all this is done in more detail.

Source file	Untitled.bas
hex file	Untitled.HEX

It should be noted that the hex file is not created until all the syntax errors the compiler can find have been eliminated by you. After a successful compilation of the code there may still be errors in the programming itself that will need to be addressed as you debug your work. Now go to the microEngineering web site (www.microEngineering Labs.com/index.htm) and download the information on the LAB-X1 experimenter board to your computer and put it in the LAB-X1 Tools folder. There are a number of very useful example programs in these files, and cutting and pasting from these to programs that you are writing will save you a lot of time. These programs are also on the support web site.

If you are familiar with and have information for the Basic Stamp, it would be a good idea to also add these files to this folder so that all your microcontroller information is in one place. If you have a CD burner on your computer (and if you do not you should get one), it is well worth your time to now copy the entire unadulterated LAB-X1 Tools folder to a CD for safekeeping. Data on a CD is much more secure than the data on a floppy drive and the best time to make a copy of it is right now before you make any changes to any of the data that you received from the vendors.

For the purposes of general discussion and experimentation, we will always call the example program that is being manipulated Untitled and the text file that is the body of the program will be called Untitled.bas.

This is the file that the compiler compiles for the microcontroller you are using to create the hex file.

The hex file that is created from this program by the compiler will be referred to as Untitled.hex. We do it this way because every time you compile and run a program the system automatically saves the program to disk at the same time. This means you lose the old program and cannot go back to it. If you are working with a complicated program, this can become a real problem because there are lots of good reasons to go back to the way things were. To avoid this pitfall, every time you load a program from disk, first save it as Untitled.bas and then play with it all you want. When you have a viable program, save it to the name that is appropriate for it. Then load the next program and change its name to Untitled.bas, and so on. I even recommend that you save each version of your program with a version designation so that you work on Blink.bas as Untitled.bas and resave it to disk as BlinkV1.0.bas; then you work on BlinkV1.0.bas as Untitled.bas and resave it as BlinkV1.1.bas, and so on. Though there is some tedium in doing this, I can assure you that it will save you a lot of headaches in the long run.

Note *The hex files created by the PBP compilers can be loaded into the PIC microcontrollers with other software/loaders. It is not imperative that hardware programmers be used.*

3

UNDERSTANDING THE MICROCHIP TECHNOLOGY PIC 16F877A: FEATURES OF THE MCU

PIC microcontrollers are manufactured by the Microchip Technology Corporation of Chandler, Arizona.

We will be using the recommended 16F877A microcontroller in the LAB-X1 board, see Figure 3.1. Not all the features provided in the 16F877A will be addressed in the exercises to follow, but enough will to give you the confidence and understanding you need to proceed on your own. In more technical terms, this MCU has the following core features (this list was reduced and modified from description by Microchip Technologies):

- High-performance RISC CPU
- Operating speed: DC-20 MHz clock input
- DC-200 ns instruction cycle
- Up to 8K × 14 words of FLASH program memory
- Up to 368 × 8 bytes of data memory (RAM)
- Up to 256 × 8 bytes of EEPROM data memory
- Interrupt capability (internal and external)
- Power-on Reset (POR)
- Power-up Timer (PWRT) and Oscillator Start-up Timer (OST)
- Watchdog Timer (WDT) with its own on-chip RC
- Programmable code-protection
- Power-saving sleep mode
- Selectable oscillator options

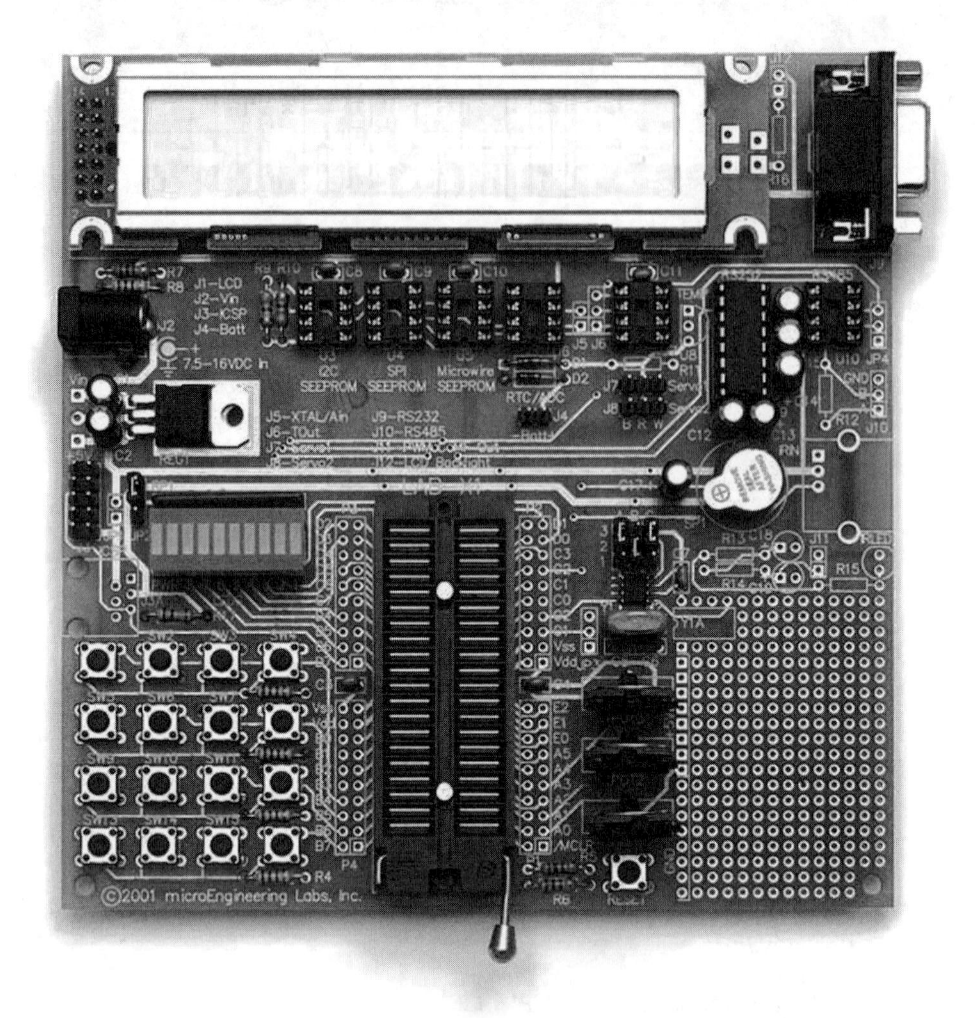

Figure 3.1 **Photograph of the LAB-X1. This image is a "very close to full size" image of the versatile Lab-X1 experimental board.**

- Low-power, high-speed CMOS FLASH/EEPROM technology
- Fully static design
- In-circuit Serial Programming (ICSP) via two pins
- Single 5V In-Circuit Serial Programming capability
- In-Circuit debugging via two pins
- Processor read/write access to program memory
- Wide operating voltage range: 2.0 to 5.5V
- High sink/source current: 25 mA
- Commercial and industrial temperature ranges
- Lowpower consumption

This MCU has the following peripheral features:

- Timer 0: 8-bit timer/counter with 8-bit pre-scaler
- Timer 1: 16-bit timer/counter with pre-scaler. It can be incremented during sleep via external crystal/clock
- Timer 2: 8-bit timer/counter with 8-bit period register, pre-scaler, and post-scaler
- Two PWM modules, maximum resolution 10 bits
- 10-bit multichannel analog-to-digital converter
- Synchronous Serial Port (SSP)
- Universal Synchronous Asynchronous Receiver Transmitter (USART)
- Parallel Slave Port (PSP), 8-bit wide
- Brown-out detection circuitry for Brown-out Reset (BOR)

This MCU is described in profuse detail in a more than two hundred-page data sheet that you can download from the Microchip website at no charge. The data sheet is a PDF document that you should have available to you at all times (maybe even open in its own window, ready for immediate access) whenever you are programming the 16F877A. The Adobe software you need to read (but not write) PDF files is also available at no charge on the web. You should have the a copy of the latest version of this very useful software on your computer.

We will not cover the entire data sheet in the exercises, but we will cover the most commonly used features of the MCU (especially the ones relevant to the LAB-X1). After doing the exercises you should be comfortable with reading the data sheet and finding the information you need to get your work done.

In this particular case, the LAB-X1 board, the MCU is already connected to the items on the board. Therefore, if you want to use the LAB-X1 for your own hardware experiments you must use the MCU pins in a way compatible with the components that are already connected to them. Often, even though the pin is being used in the LAB-X1 circuitry, you can drive something else with it without adversely affecting your experiment (depending on the load being added). Refer to Table 3.1 to quickly determine if the pin and port you want to use is free or how it is being used.

The following 40-pin PICs will work in the LAB-X1 (as of Jan 2009).

PIC16C64(A)	16C65(B)	16C662	16C67	16C74(AB)
16C765	16C77	16C774	16F74	16F747
16F77	16F777	16F871	16F874	16F874A
16F877	16F877A	16F914	16F917	18C442
18C452	18F4220	18F4320	18F4331	18F4410
18F442	18F4420	18F4431	18F4439	18F4455
18F448	18F4480	18F4510	18F4515	18F452
18F4520	18F4525	18F4539	18F4550	18F458
18F4580	18F4585	18F4610	18F4620	18F4680

TABLE 3.1 PIN DESIGNATION LISTINGS BY PORT

PORTA	PIN#	USAGE	PORT A HAS ONLY 6 EXTERNAL PINS
PORTA.0	2	5K ohm Potentiometer 0	Memory chips
PORTA.1	3	5K ohm Potentiometer 1	Memory chips
PORTA.2	4	Used by clock chips	
PORTA.3	5	5K ohm Potentiometer 2	Used by clock chips, memory
PORTA.4	6	This pin has special pull up needs!	No analog function
PORTA.5	7	Free for A to D conversion	Memory chips
PORTB			
PORTB.0	33	Keypad inputs	
PORTB.1	34	Keypad inputs	
PORTB.2	35	Keypad inputs	
PORTB.3	36	Programming device	Keypad inputs
PORTB.4	37	Keypad inputs	
PORTB.5	38	Keypad inputs	
PORTB.6	39	Programming device	Keypad inputs
PORTB.7	40	Programming device	Keypad inputs
PORTC			
PORTC.0	15	Servo/Clock	
PORTC.1	16	Clock chips, Memory chips, Servo/Clock, HPWM	
PORTC.2	17	Piezo speaker, HPWM	
PORTC.3	18	Clock chips, Memory chips, Servo/Clock	
PORTC.4	23	Used with Memory chips	
PORTC.5	24	Clock chips, A/D conversion, Memory chips	
PORTC.6	25	Transmit serial communications	RS232C
PORTC.7	26	Receive serial communications	RS232C
PORTD			
PORTD.0	19	LCD and LED bar graph	
PORTD.1	20	LCD and LED bar graph	
PORTD.2	21	LCD and LED bar graph	
PORTD.3	22	LCD and LED bar graph	

PORTD.4	27	LCD and LED bar graph
PORTD.5	28	LCD and LED bar graph
PORTD.6	29	LCD and LED bar graph
PORTD.7	30	LCD and LED bar graph
PORTE		**Port E has only 3 external pins**
PORTE.0	8	LCD writing controls
PORTE.1	9	LCD writing controls
PORTE.2	10	LCD writing controls, Communications
Other Pins		
Pin 1	MCLR	Microprocessor reset pin, pull up, Programming
Pin 11	Vdd	Logic power 5VDC, has no other use
Pin 12	Vss	Logic ground, has no other use
Pin 13	OSC1	oscillator, has no other use
Pin 14	OSC2	oscillator, has no other use
Pin 31	Vss	Logic ground, has no other use
Pin 32	Vdd	Logic power 5VDC has no other use

Re-listed in serial order, the pins are as used as listed in Table 3.2.

All the PORTB lines can be pulled up internally with a software instruction. Interrupt generation by these pins can be enabled.

TABLE 3.2 PIN DESIGNATION BY PIN NUMBER

PIN#	PIN NAME	USAGE	
Pin 1	MCLR	Processor reset pin, pull up, Programming device	
Pin 2	PORTA.0	5K ohm Potentiometer 0	
Pin 3	PORTA.1	5K ohm Potentiometer 1	
Pin 4	PORTA.2	A to D conversions	
Pin 5	PORTA.3	5K Potentiometer. Clock chips	U6
Pin 6	PORTA.4	This pin has special pull up needs! No analog function.	
Pin 7	PORTA.5	A to D conversion, Memory chips	
Pin 8	PORTE.0	LCD writing controls	
Pin 9	PORTE.1	LCD writing controls	
Pin 10	PORTE.2	LCD writing controls, Communications	

Pin 11	Vdd	Logic power, has no other use	
Pin 12	Vss	Logic ground, has no other use	
Pin 13	OSC1	Oscillator, has no other use	
Pin 14	OSC2	Oscillator, has no other use	
Pin 15	PORTC.0	Servo/Clock	
Pin 16	PORTC.1	Clock chips, Memory chips, Servo/Clock, HPWM	
Pin 17	PORTC.2	Piezo speaker, HPWM	
Pin 18	PORTC.3	Clock chips, Memory chips, Servo/Clock	
Pin 19	PORTD.0	LCD and LED bar graph	
Pin 20	PORTD.1	LCD and LED bar graph	
Pin 21	PORTD.2	LCD and LED bar graph	
Pin 22	PORTD.3	LCD and LED bar graph	
Pin 23	PORTC.4	Keypad inputs	
Pin 24	PORTC.5	Keypad inputs	
Pin 25	PORTC.6	Programming device, Keypad inputs	
Pin 26	PORTC.7	Programming device, Keypad inputs	
Pin 27	PORTD.4	LCD and LED bar graph	
Pin 28	PORTD.5	LCD and LED bar graph	
Pin 29	PORTD.6	LCD and LED bar graph	
Pin 30	PORTD.7	LCD and LED bar graph	
Pin 31	Vss	Logic ground, has no other use	
Pin 32	Vdd	Logic power, has no other use	
Pin 33	PORTB.0	Keypad inputs	
Pin 34	PORTB.1	Keypad inputs	
Pin 35	PORTB.2	Keypad inputs	
Pin 36	PORTB.3	Programming device	Keypad inputs
Pin 37	PORTB.4	Keypad inputs	
Pin 38	PORTB.5	Keypad inputs	
Pin 39	PORTB.6	Programming device	Keypad inputs
Pin 40	PORTB.7	Programming device	Keypad inputs

USING THE A TO D CAPABILITIES OF THE PIC 16F877A

A to D conversions will be discussed in more detail in Chapter 5. There are a number of basic measurements that you can make with the LAB-X1 board by using its analog-to-digital and other capabilities. These form the basis for the inputs that you can use to control the motors. The resolution of the A/D conversion can be 8 or 10 bits. Still higher resolutions are available if you use ICs that go in empty socket U6. The measurements that we make can be used to determine the following:

- Resistance
- Capacitance
- Voltage
- Frequency (not an A to D function, of course)

Resistance is measured by measuring how long it takes a resistor to discharge a capacitor that has just been charged. The measurement is as accurate as the value of the capacitor. The measurement parameters may need to be adjusted in real time to get a usable reading (meaning that the value of the two components has to be selected to get a reading in a reasonable time with reasonable accuracy).

If the relative position of the wiper on a variable potentiometer is required, the A to D conversion capabilities of the LAB-X1 can be used to read the potentiometer wiper position (not the resistance). The A to D converter always measures the voltage across the device that you connect to the analog input port. You have the choice of reading the value to a resolution of either 8 or 10 bits. If you are reading an 8 bit A to D value, the value across the resistance is divided into 256 divisions and the reading will always be between 0 and 255. If you are doing a 10-bit A to D conversion, the value will be between 0 and 1023, but since one byte can hold only 8 bits, the remaining 2 bits have to be read from another register. This is explained in greater detail in Chapter 5 in the section on setting up A to D conversions for the IC in socket U6.

THE POT COMMAND

The compiler provides the POT command to make it easy to read the resistive load placed on a pin. See the PICBASIC PRO manual for details. In order to use this command, it is necessary to set up the connection to the Lab-X1 as follows:

1. Set up the MCU for analog mode.
2. Select the pin to be used for input.
3. Select what the excitation voltage source will be (internal or external).

There are only 16 pins that may be used with the POT command, and they are the 16 pins that have been assigned the aliases from PIN0 to PIN15. These are assigned in the include file BS1DEFS.BAS. For the 16F877A, these are the pins on PORTB (0 to 7) and PORTC (8 to 15). There are different designations for the different MCUs based on the pin count. See Section 4.11 in the PICBASCIC PRO manual for a discussion of how the pin numbers are assigned for each PIC device.

The POT command is:

POT pin, scale, NBR.

The value of these variables is as follows:

- *Pin* is the pin number we have been discussing.
- *Scale* is the adjustment for various RC constants. If the RC constant is large, the value of *scale* should be small. *Scale* is determined experimentally with a potentiometer in place of the resistive load. At the low end of the resistance the value of *scale* should be 0, and at the high end it should be 255.
- *NBR* is the variable the result will be placed in.

Values between 5 and 50K ohms may be read with a 0.1 μF capacitor as shown in the Compiler manual under the POT command.

CAPACITANCE

Capacitance can be measured by determining how long it takes to charge a capacitor through an accurately calibrated resistor or by setting up an oscillator with the two components and measuring its frequency.

VOLTAGE

Voltage is measured by setting up an appropriate dividing network with precision resistors and measuring the voltage across an appropriate resistance.

FREQUENCY

The PIC 16F877A can measure frequencies directly. The timers and counters within the MCU are used to set the measurement intervals and counting hardware.

READING SWITCHES

Switches can be read from the lines of any port that is set up as an input port. Debouncing must be performed either in hardware or in software to avoid false readings. (See BUTTON command in the PICBASIC PRO manual.)

Make sure that other hardware that may be connected to the pins does not interfere with the switch function and its detection.

Reading Switches in a Matrix

Switches arranged in a matrix can be read by setting and reading the rows and columns in the matrix. The technique activates a row of buttons at a time by making it high or low and then seeing if any of the columns has been affected. A detailed description of how this is done is in Chapter 5, which discusses the keyboard of the LAB-X1 in detail.

CONFIGURING AND CONTROLLING THE PROPERTIES OF THE PORTS

The PIC 16F877A provides 33 I/O pins distributed across five ports. Each of the ports has unique capabilities built into it. This chapter discusses the capabilities of each of the ports with special attention to these special properties.

The descriptions are cursory and are designed to provide a quick and ready reference. Refer to the actual data sheet for detailed information on these ports. The data sheet provides information at a level that cannot be provided in a short introductory text like this.

PORTA

PORTA is a 6-bit wide bidirectional port with both analog and digital capability.

The general rule is that if a PIC device has any analog inputs built into it, it will come up as an analog input device on reset and startup. The PIC 16F877A has analog capability on PORTA (and PORTE), so it comes up as an analog device on startup. If you are going to use it as a digital device, you have to set register ADCON1 to %00000111. This line of code will be seen in many of the programs in this book and is explained in Chapter 9 on using LCD displays. See Table 9.8 to see how to set the various lines in PORTA and PORTE to analog or digital. (%00000111 sets all the analog pins to digital; there are many other choices.)

The PIC 16F877A supports external access to only 6 of the 8 pins on this port. Each of the 6 pins may be set to function as an input or output by appropriately loading the TRISA register. A zero in this register bit sets the corresponding pin to function as an output and a one sets it to function as an input.

Thus setting the following:

TRISA=%00111000

would make lines A0, A1, and A2 outputs and lines A3, A4, and A5 inputs. The *most significant two bits are ignored* (and could be set to 1s or 0s) because PORTA has only six active lines. (However, the two ignored bits are used by the processor and can be read when necessary. We can omit this here. See data sheet, Page 43, for specific details on how Pins 6 and 7 are used by the in circuit debugger.)

> **Note** *The % symbol means that this is a binary number. We will use this binary notation throughout the book because it makes it easier to see what each bit is being set to. Bit 7, the most significant bit, is on the left, bit 0 the least significant is on the right).*

The specific functions of the pins are controlled with the ADCON1 (the first A to D control) register.

All the pins have TTL level inputs and full CMOS level output drivers. This makes it easy to connect these lines directly to standard logic components, meaning that *usually* no intermediate resistors are needed between components if TTL or CMOS components are connected.

PORTA designations are somewhat complicated. Pins A0, A1, A2, A3 (skip A4), and Pin A5 can be configured as analog inputs by setting the ADCON1 register. Pin A3 is also used as a voltage input for comparing with the analog voltage inputs. Pin PORT A4 is used for the TIMER0 input and is then called T0CK1. This is Pin 6 of the PIC, and it is used as the input pin for TIMER0 only when configured as such. It is a Schmitt-triggered input with open drain output. Open drain means that it acts like the contacts of a tiny relay that go to 0 volts when closed but float when open and not connected to anything. Schmitt-triggered inputs have increased noise immunity.

The two registers that control PORTA are TRISA and ADCON1.

ADCON1 controls the A to D and voltage reference functions or PORTA. The setting of the various bits selects a complicated set of conditions that are described in detail in Table 9.8. (In the preceding discussion when ADCON1 was set to %00000111, we were accessing this feature.)

Pin A4 has special needs when used as an output. It can be pulled down low but will float when set high. It must be pulled up with a (10K to 100K) resistor to tie it high. This pin has an open drain output rather than the usual bipolar state of the other pins. This pin is skipped in the A to D conversion table.

PORTB

PORTB is a full 8-bit wide bidirectional port.

Internal circuitry (built into the MCU) allows all the pins on PORTB to be pulled up to a high state (very weakly) by setting pin 7 of the Option Register (OPTION_REG.7) to 0. These pull-ups are disabled on startup and on reset.

Pins B3, B6, and B7 are used for the low voltage programming of the PIC. Bit 3 in TRISB must be cleared (set to 0 or pulled down to 0) to negate the pull-up on this pin to allow programming to take place. See Pages 42 and 142 in the data sheet for more information on the B3 pin. It is important to keep this in mind because if for any reason Pin B3 cannot be made low it will not be possible to program the device.

Pins B4 to B7 will cause an interrupt to occur when their state changes if they are configured as inputs and the appropriate interrupts are configured. Pins that are configured as outputs will be excluded from the interrupt feature. The interrupts are controlled by the INTCON (Interrupt Control) register. This PORTB interrupt capability has the special feature that it can be used to awaken a sleeping MCU.

Pin B0 has separate (external) interrupt functions that are controlled through the INTEDG bit which is bit 6 of the OPTION_REG. (See the data sheet for more information.) External interrupts are routed to the PIC through this pin.

The three registers that control PORTB are TRISB, INTCON, and the OPTION_REG.

OPTION_REG controls the optional functions of PORTB as follows:

- Bit 7 of OPTION_REG sets the pull-ups; programming uses
- Bit 6 of OPTION_REG sets edge selection for interrupts; programming uses
- Bit 5 of OPTION_REG sets the clock selection
- Bit 4 of OPTION_REG sets Timer 0 input pulse edge condition
- Bit 3 of OPTION_REG sets the pre-scaler option; used in low voltage programming
- Bit 2 of OPTION_REG sets pre-scaler value
- Bit 1 of OPTION_REG sets pre-scaler value
- Bit 0 of OPTION_REG sets pre-scaler value

PORTC

PORTC is a full 8-bit wide bidirectional port.

All the pins on PORTC have Schmitt-trigger input buffers. This means that they are designed to be more immune to noise on the input lines.

The alternate functions of the PORTC pins are defined as follows:

- Pin C0 I/O pin or Timer1 oscillator output or Timer1 Clock input
- Pin C1 I/O pin or Timer1 oscillator input or Capture 2 input or Compare 2 output or Hardware PWM2 output
- Pin C2 I/O pin or Capture 1 input or Compare 1 output or Hardware PWM1 output
- Pin C3 I/O pin or Synchronous clock for both SPI and I2C memory modes
- Pin C4 I/O pin or SPI data or data I/O for I2C mode
- Pin C5 I/O pin or Synchronous serial port data output
- Pin C6 I/O pin or USART Asynchronous transmit or synchronous clock
- Pin C7 I/O pin or USART Asynchronous receive or synchronous data

Special care has to be taken when using PORTC's special function capabilities in that certain of these functions will change or set the I/O status of certain other pins when in use, and this can cause unforeseen complications in the function of other capabilities. See the data sheet for details.

The register that controls PORTC is the TRISC register. No other registers are involved. DEFINEs are used to control certain functions. (Using the DEFINEs is covered in the PROBASIC PRO compiler language manual. This is the manual for the language we will be using to program the PIC 16F877A. The manual is provided, as a part of the compiler documentation, by microEngineering Labs.)

The speaker on the LAB-X1 board is connected to pin C1, so the use of this pin is limited because the noise generated by the speaker when this pin is used can be very irritating. Since this is one of the lines that allows the generation of continuous background PWM signals (HPWM 2), it compromises the clean use of this pin unless the speaker is removed. However, I recommend that you avoid modifications to the board if you can. The load of the tiny speaker loads the pin and can compromise a few other uses but is okay for most uses.

PORTD

PORTD is a full 8-bit wide bidirectional port.

All the pins on PORTD have Schmitt-trigger input buffers. This means that they are designed to be more immune to noise on the input lines.

PORTD can also be configured as a microprocessor port by setting PSPMODE TRISE.4 to 1. (Note that you are specifying bit 4 of PORTE here internally; there is no external pin 4.) In this mode all the input pins are in TTL mode.

The alternate function of the PORTD pins are defined as follows:

- Pin D0 or parallel slave port bit 0
- Pin D1 or parallel slave port bit 1
- Pin D2 or parallel slave port bit 2
- Pin D3 or parallel slave port bit 3
- Pin D4 or parallel slave port bit 4
- Pin D5 or parallel slave port bit 5
- Pin D6 or parallel slave port bit 6
- Pin D7 or parallel slave port bit 7

The registers that control PORTD are the TRISD register and the TRISE register. TRISE controls the operation of the PORTD parallel slave port mode when Bit PORTE.4 is set to 1. (Again, only pins E0, E1, and E2 are available external to the MCU on PORTE.)

Slave port functions as set by PORTE when Bit PORTE.4 is set to 1 are as follows:

- Bit TRISE.0 direction control of Pin PORTE.0 / RD / AN5
- Bit TRISE.1 direction control of Pin PORTE.1 / WR / AN6
- Bit TRISE.2 direction control of Pin PORTE.2 / CS / AN7
- Bit TRISE.3. NOT USED
- Bit TRISE.4 Slave port select, 1 = Port selected, 0 = Use as standard I/O port
- Bit TRISE.5 Buffer overflow detect, 1 = Write occurred before reading old data, 0 = No error occurred
- Bit TRISE.6 Buffer status, 1 = still holds word, 0 = has been read
- Bit TRISE.7 Input buffer status, 1 = full, 0 = nothing received

Read the data sheets to get a better understanding of these operations. The preceding list is a very quick overview and is intended only to alert you and to give you an idea of what the possibilities are.

PORTE

PORTE is only three external bits wide and is a bidirectional port. The other bits are internal and are used as mentioned in the PORTD section (to which they are related). The pins on PORTE can be configured as analog or digital.

All the pins on PORTE have Schmitt-trigger input buffers.

The alternate function of the PORTE pins are defined as follows:

- Pin RE0 direction control of Pin PORTE.0 / RD / AN5
- Pin RE1 direction control of Pin PORTE.1 / WR / AN6
- Pin RE2 direction control of Pin PORTE.2 / CS / AN7

TIMERS

The three timers in the PIC 16F877A allow the accurate timing and counting of chronological events. Timers are discussed is much greater detail in Chapter 6, which is devoted exclusively to timers and counters. A fourth timer provides a watchdog function. Each timer occupies a 1- or 2-byte location in the memory.

Some of the timers have pre-scalers associated with them that can be used to multiply the timer setting by an integer amount. As you can imagine, the scaling ability is not adequate to allow all exact time intervals to be created. You also have to consider the uncertainty in the frequency of the clocking crystal, which is usually not exactly what it is stated to be and may drift with its temperature. This means that though fairly accurate timings can be achieved with the hardware as received, additional software adjustments may have to be added if more accurate results are desired. You do this by having the software make a correction to the timing every so often. (This also means that an external source that is at least as accurate as the result you want is needed to verify the timing accuracy of the device created.)

The three timers in the microcontroller are clocked at a fourth of the oscillator speed, meaning that a timer using a 4 MHz clock gets a counting signal at 1 MHz.

Very simply stated, an 8-bit timer will count from 0 up to 255 and then flip to 0 and start counting from 0 to 255 again. An interrupt occurs every time the timer registers overflows from 255 to 0. You respond to the interrupt by doing whatever needs to be done and then resetting the interrupt flag. On timers that permit the use of a prescalar, the prescalar allows you to increase the time between interrupts by multiplying the time between interrupts with a definable value in a 2-, 3-, or 4-bit location. On timers that can be written to, you can start the counter wherever you like to change the interrupt timing; on timers that can be read, you can read the contents whenever you like. For example, a 1-second timer setting with a prescalar set to 16 would provide you with an interrupt every 16 seconds. You will have 16 seconds to do whatever you wanted to do between the interrupts before you will miss the next interrupt.

If you needed an interrupt every 14.5 seconds, you would use a timer set to 0.5 seconds and a prescalar of 29, if 29 was specifiable (which it is not here). So not all time intervals can be created with this strategy because there are limits as to what can be put in the timer and what can be put in the prescalar when you are using 8-bit registers and specific oscillator speeds.

Pre-scalers The value of the scaling factor that will be applied to the timer is determined by the contents of 2 or 3 bits in the interrupt control register. These bits multiply the time between interrupts by powers of 2 as explained in Chapter 6. Pre-scalers and post-scalers have the same effect on the interrupts: they delay them.

Watchdog Timer A watchdog timer sets an interrupt when it runs out to tell you that for some reason the program has hung up or otherwise gone awry. As such, it is expected that in a properly written program the watchdog timer will never set an interrupt. This is accomplished by resetting the watchdog timer every so often within the program. The compiler does this automatically if the watchdog timer option is set. Setting the option does not guarantee a program that cannot hang up. Software errors and infinite loops that reset the timer within them can still cause hangups.

Counters Both Timer0 and Timer1 can be used as counters. Timer2 cannot be used as a counter because it has no internal or external input pin. The timers and the counters are covered in detail in Chapter 6.

The following address and web sites may be used to contact Microchip Technologies. The website provides downloads for the data sheets.

Microchip Technology Corporation, Inc.
2355 West Chandler Boulevard
Chandler, Arizona 85224-6199
phone: (480) 792-7200
fax: (480) 899-9210
Website: www.microchip.com

microEngineering Labs maintains a very useful and helpful web site that will also be a tremendous aid to you as you learn about the PIC microcontrollers by using their LAB-X1. They can be reached at:

microEngineering Labs Inc.
Box 60039
Colorado Springs, CO 80960-0039

or:

microEngineering Labs
1750 Brantfeather Grove
Colorado Springs, CO 80960
phone: (719) 520-5253
fax: (719) 520-1867
e-mail: support@microengineeringlabs.com
Website: www.microengineeringlabs.com/index.htm

4

THE SOFTWARE, COMPILERS, AND EDITORS

microEngineering Labs provides two BASIC compilers that make writing the code for the PIC family of microcontrollers provided by the Microchip Corporation tremendously easier than it would otherwise be. We will discuss the more powerful of the two, the PICBASIC PRO Compiler, only. A listing of the commands provided by each compiler is provided in the next section to allow you to compare the two compilers and select the one best suited to your needs.

Basic Compiler Instruction Set

The following is a list of the commands provided in the smaller compiler:

ASM..ENDASM	Insert assembly language code chapter
BRANCH	Computed GOTO (equivalent to onGOTO)
BUTTON	Debounce and auto-repeat input on specified pin
CALL	Call assembly language subroutine
EEPROM	Define initial contents of on-chip EEPROM
END	Stop execution and enter low power mode
FOR..NEXT	Repeatedly execute statement(s)
GOSUB	Call BASIC subroutine at specified label
GOTO	Continue execution at specified label
HIGH	Make pin output high
I2CIN	Read bytes from I2C device

I2COUT	Send bytes to I2C device
IF..THEN-GOTO	If specified condition is true
INPUT	Make pin an input
[LET]	Assign result of an expression to a variable
LOOKDOWN	Search table for value
LOOKUP	Fetch value from table
LOW	Make pin output low
NAP	Power down processor for short period of time
OUTPUT	Make pin an output
PAUSE	Delay (1 mSec resolution)
PEEK	Read byte from register
POKE	Write byte to register
POT	Read potentiometer on specified pin
PULSIN	Measure pulse width (10 μs resolution)
PULSOUT	Generate pulse (10 μs resolution)
PWM	Output pulse width modulated pulse train to pin
RANDOM	Generate pseudo-random number
READ	Read byte from on-chip EEPROM
RETURN	Continue execution at statement following last executed GOSUB
REVERSE	Make output pin an input or an input pin an output
SERIN	Asynchronous serial input (8N1)
SEROUT	Asynchronous serial output (8N1)
SLEEP	Power down processor for a period of time (1 second resolution)
SOUND	Generate tone or white-noise on specified pin
TOGGLE	Make pin output and toggle state
WRITE	Write byte to on-chip EEPROM

MATH OPERATIONS

All math operations are unsigned and performed with 16-bit precision:

+	(Addition)
–	(Subtraction)
*	(Multiplication)
**	(MSB of Multiplication)
/	(Division)

//	(Remainder)
MIN	(Minimum)
MAX	(Maximum)
&	(Bitwise AND)
|	(Bitwise OR)
^	(Bitwise XOR)
&/	(Bitwise AND NOT)
|/	(Bitwise OR NOT)
^/	(Bitwise XOR NOT)

PICBASIC PRO Compiler Instruction Set

Here is a list of the commands provided in the larger compiler.

@	Insert one line of assembly language code
ADCIN	Read on-chip analog to digital converter
ASM..ENDASM	Insert assembly language code chapter
BRANCH	Computed GOTO [equivalent to ON..GOTO]
BRANCHL	Branch out of page [long BRANCH]
BUTTON	Debounce and auto-repeat input on specified pin
CALL	Call assembly language subroutine
CLEAR	Zero all variables
CLEARWDT	Clear [tickle] Watchdog Timer
COUNT	Count number of pulses on a pin
DATA	Define initial contents of on-chip EEPROM
DEBUG	Asynchronous serial output to fixed pin and baud
DEBUGIN	Asynchronous serial input from fixed pin and baud
DISABLE	Disable ON DEBUG and ON INTERRUPT processing
DISABLE DEBUG	Disable ON DEBUG processing
DISABLE INTERRUPT	Disable ON INTERRUPT processing
DTMFOUT	Produce touchtones on a pin
EEPROM	Define initial contents of on-chip EEPROM
ENABLE	Enable ON DEBUG and ON INTERRUPT processing
ENABLE DEBUG	Enable ON DEBUG processing
ENABLE INTERRUPT	Enable ON INTERRUPT processing

END	Stop execution and enter low power mode
ERASECODE	Erase block of code memory
FOR ..NEXT	Repeatedly execute statements
FREQOUT	Produce up to two frequencies on a pin
GOSUB	Call BASIC subroutine at specified label
GOTO	Continue execution at specified label
HIGH	Make pin output high
HPWM	Output hardware pulse width modulated pulse train
HSERIN	Hardware asynchronous serial input
HSERIN2	Hardware asynchronous serial input, second port
HSEROUT	Hardware asynchronous serial output
HSEROUT2	Hardware asynchronous serial output, second port
I2CREAD	Read from I2C device
I2CWRITE	Write to I2C device
IF..THEN..ELSE..ENDIF	Conditionally execute statements
INPUT	Make pin an input
LCDIN	Read from LCD RAM
LCDOUT	Display characters on LCD
{LET}	Assign result of an expression to a variable
LOOKDOWN	Search constant table for value
LOOKDOWN2	Search constant/variable table for value
LOOKUP	Fetch constant value from table
LOOKUP2	Fetch constant/variable value from table
LOW	Make pin output low
NAP	Power down processor for short period of time
ON DEBUG	Execute BASIC debug monitor
ON INTERRUPT	Execute BASIC subroutine on an interrupt
OWIN	One-wire input
OWOUT	One-wire output
OUTPUT	Make pin an output
PAUSE	Delay, 1 ms resolution
PAUSEUS	Delay, 1 µs resolution
PEEK	Read byte from register
PEEKCODE	Read byte from code space

POKE	Write byte to register
POKECODE	Write byte to code space at device programming time
POT	Read potentiometer on specified pin
PULSIN	Measure pulse width on a pin
PULSOUT	Generate pulse to a pin
PWM	Output pulse width modulated pulse train to pin
RANDOM	Generate pseudo-random number
RCTIME	Measure pulse width on a pin
READ	Read byte from on-chip EEPROM
READCODE	Read word from code memory
REPEAT..UNTIL	Execute statements until condition is true
RESUME	Continue execution after interrupt handling
RETURN	Continue at statement following last GOSUB
REVERSE	Make output pin an input or an input pin an output
SELECT CASE	Compare a variable with different values
SERIN	Asynchronous serial input, BS1 style
SERIN2	Asynchronous serial input, BS2 style
SEROUT	Asynchronous serial output, BS1 style
SEROUT2	Asynchronous serial output, BS2 style
SHIFTIN	Synchronous serial input
SHIFTOUT	Synchronous serial output
SLEEP	Power down processor for a period of time
SOUND	Generate tone or white noise on specified pin
STOP	Stop program execution
SWAP	Exchange the values of two variables
TOGGLE	Make pin output and toggle state
USBIN	USB input
USBINIT	Initialize USB
USBOUT	USB output
WHILE..WEND	Execute statements while condition is true
WRITE	Write byte to on-chip EEPROM
WRITECODE	Write word to code memory
XIN	X-10 input
XOUT	X-10 output

MATH FUNCTIONS AND OPERATORS

The math operations are unsigned and performed with 16-bit precision.

+ (Addition)
– (Subtraction)
* (Multiplication)
** (Top 16 bits of multiplication)
*/ (Middle 16 bits of multiplication)
/ (Division)
// (Remainder [modulus])
<< (Shift Left)
>> (Shift Right)
ABS (Absolute Value)
COS (Cosine)
DCD (2n Decode)
DIG (Digit)
DIV32 (31-bit × 15-bit Divide)
MAX (Maximum)
MIN (Minimum)
NCD (Encode)
REV (Reverse Bits)
SIN (Sine)
SQR (Square Root)
& (Bitwise AND)
| (Bitwise OR)
^ (Bitwise Exclusive OR)
~ (Bitwise NOT)
&/ (Bitwise NOT AND)
|/ (Bitwise NOT OR)
^/ (Bitwise NOT Exclusive OR)

As you can see from the preceding comparison, the PICBASIC PRO compiler provides a much more comprehensive instruction set and is the compiler of choice for serious development work. The mathematical functions are substantially more powerful.

It is, of course, also possible to program microcontrollers in assembly language and "C" (and other languages), but this book does not cover this programming. There are a number of good books on the subject and some that I looked over are listed in a file on the support web site for this book (www.encodergeek.com) with my comments.

Some educators feel that a junior college-level class on the subject is the best way to learn how to do this and there is some merit to this but for our purposes the PICBASIC PRO compiler will do everything we need.

In addition to the compiler, you need an editor to allow you to write and edit programs with ease. A very adequate editor, the MicroCode Studio editor, is provided as a part of the compiler package. This comprehensive and powerful editor is available on the Internet at no charge from MicroCode Studios. This is a complete editor with no limit on the number of lines of code that you can write. It is fully integrated with the software and hardware provided by microEngineering Labs and is the editor of choice for most users. The free version is limited to compiling programs for just a few microcontrollers, but these include both the 16F877A, 18F4331, and the 16F84A.

The editors available are as follows:

- **MicroCode Studio** Mecanique's MicroCode Studio is a powerful, visual Integrated Development Environment (IDE) with In-Circuit Debugging (ICD) capability designed specifically for microEngineering Labs PICBASIC PRO compiler. This software can be downloaded from the Internet at no charge. The only limitation on it is that it allows you to run only one IDE at one time, but that's not a real handicap at our level of interest. This is the editor that best suits our needs and all programs in this book were written with this editor.
- **Proton+** This is a lite basic editor provided by CrownHill. This is a test version of their editor and is limited to 50 lines of code and three processors including the PIC 16F877A. It's a nice editor but limited in the free version to the 50 lines of code. If you like this editor, you can use this as your main editor and then cut and paste to the MicroCode Studio to compile and run your programs and thus go around the 50-line limit. The native language of this editor is not the same as PicBasic so there are other handicaps to contend with, which are best avoided.
- **MicroChip MPLAB** This is the software that the maker of the microcontrollers, Microchip Technologies, provided for editing programs written for their PIC series of microcontrollers. It is an assembly level programmer. We are not going to be doing any assembly language programming, but the editor can be useful and you should be aware of its existence.

PICBASIC PRO Compiler

The PICBASIC PRO Compiler (referred to as the PBP hereafter) provides all the functions needed to program almost the entire family of PIC microcontrollers in a BASIC-like environment. This means that it allows you to write programs that read the inputs and write to the outputs in a simple and easy to learn way. It means that communications are simplified and the time it takes to get an application running is reduced many fold. It means that the programs are easier to follow and to debug (though debugging can get quite complicated even on these seemingly simple devices). The compiler supports only integer math, but that is not a big handicap when we are working with these limited microprocessors. You need to select a much more powerful microprocessor if mathematics is a major need for your application. In our particular case, where we are running encoded DC motors, this inhibits the implementation of the differential function in the PID loop, as you will see later on.

It also means that the programs that are developed are longer than assembly language programs and slower in their execution than assembly language programs. There are also complications that have to do with the use of interrupts that have to be addressed, but these are beyond the scope of this book.

All the exercises and examples provided in the text are based on the PBP compiler. We will not go over the detailed instructions for using each of the PBP instructions in the text. It is expected and will be assumed that you will have purchased the software and thus will have the manual for the compiler in hand. However, there are some commands that can be complicated to implement and we will spend time on these.

The compiler is kept current by microEngineering Labs for the latest MCUs released by the Microchip Technologies Corporation. The LAB-X1 uses the 16F877A MCU, and it is the MCU of choice though other MCUs that have a general pin-for-pin compatibility with this MCU may also be used. All the experiments and exercises in this book will use the PIC 16F877A only. The compiler addresses almost all the capabilities of this MCU, and we will cover the use of all the devices that are provided on the LAB-X1 board. (The 18F4331 is used for the encoder attached motors only.)

Detailed instructions for installing the software on your PC are provided in the compiler book. It is not necessary to install the software from a DOS prompt. It is much easier to install it under Windows with the Install.exe or equivalent file provided in each package.

The software can be set up so that one mouse click will transfer the program from the editor to the PIC microcontroller and run the program in the PIC. In order to do this you have to add a couple of functional codes to the programmer operating system. These codes tell the programmer to load the program and execute it. Installing the software was covered in detail in Chapter 2.

A SIMPLE EXAMPLE PROGRAM USING PIC BASIC

A program that makes the LEDs blink on and off is usually the first program written by beginners. The purpose of the program is not to blink the LEDs but rather to allow you to go through the programming procedures in a simple and straightforward way and get a result that is easy to verify. Once you have the LEDs blinking, you will know that you have followed all the steps necessary to write and execute a program. Larger, more complicated programs may be much more difficult to write and debug, but they are no more difficult to compile, load, and run.

Following are the keystrokes for writing and running the Blink the LEDs program:

Program 4.1 **The First Program.** Blinking all 8 LEDs on PORT D one at a time

```
; ***********************************************************
; * Name       myBlink8leds.BAS
; * Author     Harprit Singh Sandhu
; * Notice     Copyright (c) 2008
```

(continued)

Program 4.1 The First Program. Blinking all 8 LEDs on PORT D one at a time (*continued*)

```
; *                 All Rights Reserved
; * Date            1/Feb/2008
; * Version         1.0
; * Notes           Blinks all 8 LEDs on bargraph one at a time
; ***************************************************************
CLEAR                       ; clear all memory
DEFINE OSC 4                ; define the osc freq
LED_ID VAR BYTE             ; call out the two variables LED_ID
                            ; and I
I VAR BYTE                  ; as 8 bit bytes
TRISD = %00000000           ; set PORTD to all outputs
                            ;
MAINLOOP:                   ; loop is executed forever
   I = 1                    ; initialize the counter to 1
   FOR LED_ID = 1 TO 8      ; do it for the 8 LEDs
     PORTD = I              ; puts number in PORTD
     PAUSE 100              ; pause so you can see the display
     I = I * 2              ; multiplying by 2 moves lit
                            ; LED left 1 pos
   NEXT LED_ID              ; go up and increment counter
GOTO MAINLOOP               ; Do it all forever
END                         ; always end with END statement
```

PICBASIC PRO Tips and Cautions

1. To get context sensitive help, move the cursor over a PICBASIC command, click to set cursor and press F1.
2. All the programs assume the PIC is running at 4 MHz. To change the default setting (for example, to 20 MHz), simply add DEFINE OSC 20 at the top of your program and set the jumpers on the LAB-X1 accordingly. It is good practice to always specify the oscillator speed in a program. Beginners should start with 4 MHz designs. The LAB-X1 is set up to run at 4 MHz as received from the factory. See the PBP manual for further details of assumptions and conventions used by the software. The defined OSC speed has to match the hardware crystal for the software to work correctly.
3. Before you can use the LCDisplay on the LAB-X1, ADCON1 must be set (to %00000111) and you must pause about 500 ms to allow the LCD to start up before issuing the first command. You may not need a pause, or a shorter pause may be specified if there are a lot of time consuming instructions before the first LCDOUT instruction is executed. (Other values of ADCON1 can also be used depending on how you want the A and E ports configured. See discussion in Chapter 9 on using the LCD.)
4. I have used binary notation (%01010101) throughout the book to set relevant bytes and registers so that you can readily see which bit is being set to what. The compiler

accepts hexadecimal and decimal notation just as willingly. Binary notation does not permit a space after the % sign and all eight bits must be specified.

5. A single quotation mark (') when copied from a Word file and pasted into the MicroCode Studio editor will be interpreted as a (`) and will therefore not properly start the comment part of the line. All these have to be changed in the editor after pasting. Pasting from the editor into Word does not exhibit the same effect. If you use a semicolon (;) for the comments, this problem does not occur.
6. All the named registers can be called by name when using the compilers. The register names are the same as those used (defined) by the manufacturer in the data sheets and are the same across the entire family of PIC microcontrollers if they provide the same function. Uppercase or lowercase names can be used. The DEFINEs must be stated in uppercase only and the spellings in the DEFINE lines are not always checked by the compiler! Be very careful when entering DEFINEs into your program.
7. Circuits and segments of circuits are provided throughout this book to show you how to connect up to the hardware when you design your own circuits. If you have access to AutoCAD you can use the diagrams in the files on the support web site to cut and paste into your own designs.

A FREE DEMO BASIC COMPILER

A free version of the PIC Basic Pro compiler by microEngineering Labs can also be downloaded from the microEngineering Labs web site. This is a fully functional compiler with the limitation that programs are limited to 30 lines of code. This is enough to allow you to test the compiler and any instruction that you might have a special interest in. This version can give you a good idea of the power and ease of use of the language. Try it.

5

CONTROLLING THE OUTPUT AND READING THE INPUT

In this chapter we will learn how we interface an MCU to the real world by first learning how to create outputs with the microprocessor and then learning how to read inputs into the microprocessor. In following chapters we will combine the outputs and the inputs to control the operation of small motors of all kinds.

All the programs that we will be discussing are provided on the support web site for this book. You can copy them from the site to run them. The exercises listed at the end of various chapters are designed to increase your familiarity and competence with the 16F877A. The answers to them are not provided.

In preparation for writing programs, set up the LAB-X1 so that it can be programmed with one mouse button click or by pressing F10 as is described in detail in Appendix A.

The I/O that uses ICs in the seven empty sockets on the LAB-X1 board is covered separately in Chapters 7 and 8. These chapters also cover one wire memory, A to D converters, and a number of thermometric devices.

The I/O that uses the serial port (as RS232 or RS485) is covered in Chapter 8. Specifically, this covers communications between the PIC 16F877A and personal computers.

We will learn about input and output by writing simple programs that control the outputs and read the inputs. We will learn how to control the outputs first because this can be done directly from the software without need for any input or any external hardware. Once we can control the output, we will learn how to read the inputs and make them interact with the output.

The following are output programs to be developed:

- A program to blink one LED on the bar graph.
- Blink all eight LEDs in the bar graph consecutively.

- Dim and brighten one LED.
- Write "Hello World" to the LCD on its two lines.
- Write binary and decimal values to the LCD.
- Output a simple tone on the speaker.
- Output a telephone tone signal on the speaker.
- Advanced: Move an R/C servo back and forth.

The following are input programs to be developed:

- A program to read the first column, first row button, and turn on one LED while this button is down.
- Read the entire keyboard and display the binary value of the row and column read on the LCD.
- Read the keyboard and display decimal key number on the LCD.
- Read one potentiometer and display its 8-bit value on the LCD in binary, hex, and decimal notation. Also display the binary value on the bar graph.
- Read all three potentiometers on the LAB-X1 and display their values on the LCD.
- Advanced: Use the three potentiometers on the LAB-X1 to control an R/C servo. Control the location of the center position, the limit position of the end positions, and the rate of movement. Use three switches on the keypad to move the servo clockwise, center the servo, and move it counterclockwise.

Generating Outputs

It will be easier if we learn to control the outputs first because we can do this from programs that we write without the need for any additional hardware or input signal. We will start with the simple control of LEDs and proceed to the control of the two-line LCD that is provided on the LAB-X1, and then move on to using the speaker and an R/C hobby servo.

Let us start with the standard turning an LED on and off program. We will use one of the LEDs in the ten-LED bar graph that is provided on the LAB-X1. On the LAB-X1 we have control of only the rightmost eight LEDs on the bar graph. The leftmost LED is the power-on indicator and the one next to it comes on if we were using a common cathode arrangement (as opposed to the common anode arrangement as it is currently configured).

The circuitry we are interested in is shown in Figure 5.1. All other circuitry of the LAB-X1 is still in place, but we have suppressed it, as shown in the figure, so we won't be distracted by it and can concentrate on the one LED that is of interest, PORTD.0. (PORTD.0 refers to bit 0 of PORTD.)

This is how we turn something on with a microprocessor. We will use this technique whenever we need to turn something on in our experiments. If the signal needs to be

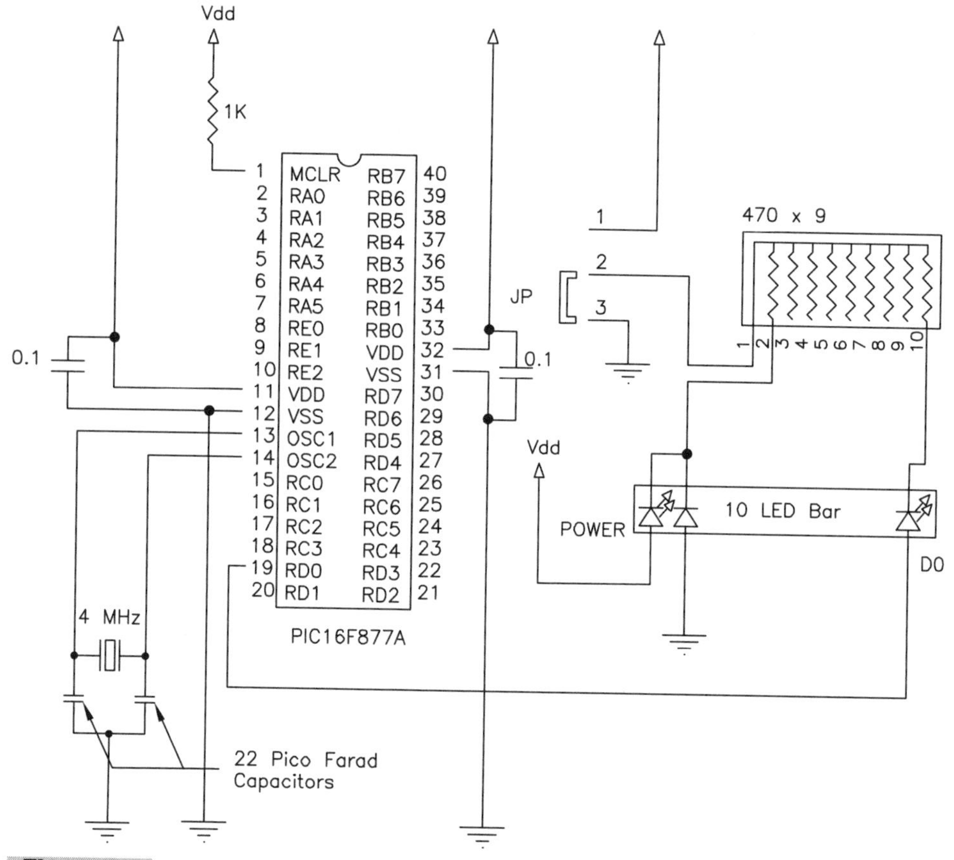

Figure 5.1 The LED bar graph circuitry to PORTD pin 0

amplified to do useful work, we will do that. Transistors, conventional relays, and solid state relays can all be controlled by TTL level signals to give us the control voltages and amperages we need.

The following paragraphs and Program 5.1 guide through your first interaction with the microcontroller. We will take all the steps necessary to write an operational program and run it on the LAB-X1. Though this is a very simple program the steps taken here will be repeated for all the programs that we will ever write. It is important that you understand each and every step undertaken here before we proceed any further.

In this first experiment, we want to control the rightmost LED of the LED array. This is connected to bit 0 of PORTD in the circuitry shown in Figure 5.1. Our program needs to turn this LED on and off to demonstrate that we have control of these two functions.

In general, the ports on the microcontrollers (MCUs) are designed so that they can be used as inputs or outputs. In fact, the ports can be programmed so that certain pins on a port are inputs and others are outputs. All we have to do is tell the program what we want done and the compiler will handle the details. The compiler not only allows you to define how you will use the pins of each port, it can also set them up as inputs or as outputs automatically, depending on the instructions that we use in our programs. You have a choice of setting PORTD to an output port and then setting pin 1 on this port high, or you can simply tell the compiler to make pin 1 of PORTD high and it will take care of the details.

The ports can be treated just like any other memory location in the microcontroller. By name, you can read them, set them, and use them in calculations and manipulations just like you can with any other named or unnamed memory location. If things are connected to the ports and pins, the program will interact with and respond to whatever is connected to them. (Any named port, register, or pin can be addressed directly by name for all purposes when using the PBP Compiler. They are called out as they are named in the data sheet.)

BLINK ONE LED

Type Program 5.1 as follows into your PC and save it. It does not need to be saved in the same directory as the PBP.exe program. To keep the conventions being used in the compiler manual, call this program myBLINKL so that it does not overwrite the BLINK.BAS program provided on the disk that came with the LAB-X1. Program 5.1 here demonstrates the on-off control pin 0 of PORT D.

Program 5.1 Controlling (blinking) an LED. Blinks the rightmost LED on bar graph

```
CLEAR                 ; clear memory locations
DEFINE OSC 4          ; osc speed. We will use 4 MHz
                      ; for all
                      ; our initial experiments
LOOP:                 ; main loop
  HIGH PORTD.0        ; turns LED connected to D0 on
  PAUSE 500           ; delay 0.5 seconds
  LOW PORTD.0         ; turns LED connected to D0 off
  PAUSE 500           ; delay 0.5 seconds
GOTO LOOP             ; go back to Loop and repeat operation
END                   ; all programs must end with END
```

The program demonstrates the most elementary control we have over an output. In this program we did not have to set the port directions (with the TRIS command) because the HIGH and LOW commands take care of that automatically. (If we used PORTD.0=1 instead of HIGH PORTD.0 we would have to set TRISD to %11111110 first to set all lines to inputs except D0, which is here shown set as an output.)

We will use binary notation (%11110000) for setting all ports and port directions throughout this book, though you can use hexadecimal ($F0) and decimal (DEC 240) notation interchangeably. Using binary notation lets you see what each pin is doing without having to make any mental conversions.

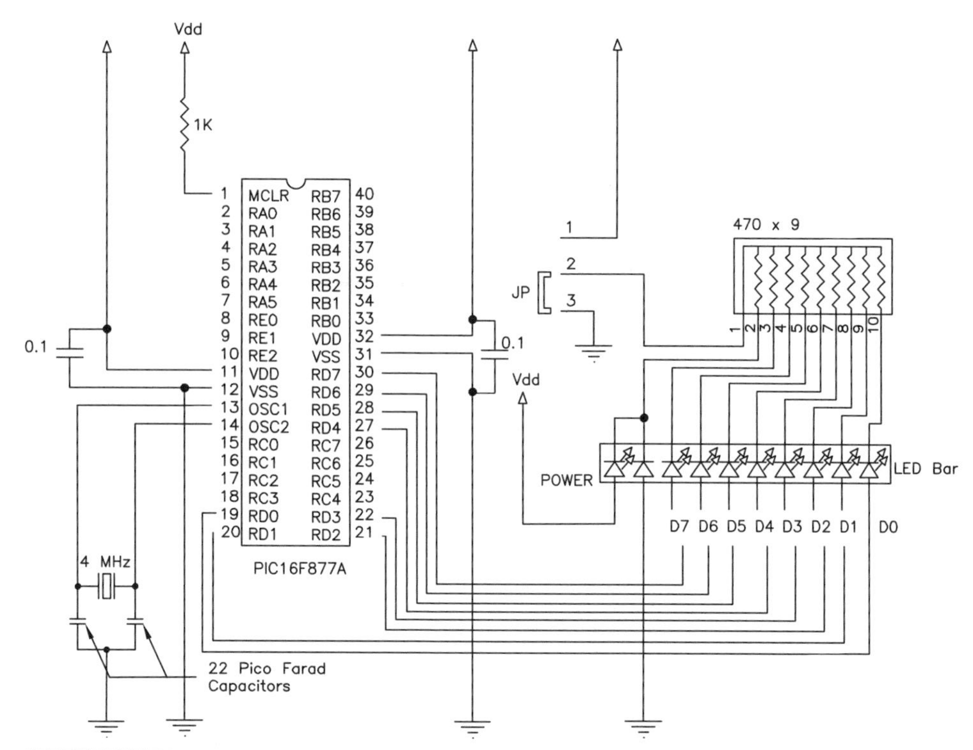

Figure 5.2 The LED bar graph circuitry to all of PORTD.

BLINK EIGHT LEDS IN SEQUENCE

In the next experiment (Program 5.2), the circuitry for which is shown in Figure 5.2, we will blink the eight rightmost LEDs on the bar graph one LED at a time. We do this by setting PORTD to 1 and then multiplying it by two eight times to move the lighted LED left in each iteration. Note that the last multiplication overflows the 8-bit counter and turns all the LEDs off.

Program 5.2 Blinking 8 LEDs one after the other on bar graph

```
CLEAR                     ; clear memory
DEFINE OSC 4              ; osc speed
LEDID VAR BYTE            ; call out the two variables
A VAR BYTE                ; as 8 bit bytes
TRISD = %00000000         ; set PORTD to all outputs
                          ;
MAINLOOP:                 ; this loop is executed forever
  A = 1                   ; initialize the counter to 1
    FOR LEDID = 1 TO 8 ; do it for the 8 LEDs
      PORTD = A           ; puts number in PORTD
      PAUSE 100           ; pause so you can see the display
```

(continued)

Program 5.2 **Blinking 8 LEDs one after the other on bar graph** (*continued*)

```
        A = A * 2           ; multiply by 2 moves lit LED left 1
                            ; position
      NEXT LEDID            ; go up and increment counter
  GOTO MAINLOOP             ; do it all forever
  END                       ; always end with END
```

DIMMING AND BRIGHTENING ONE LED

In Program 5.3 we demonstrate the ability to dim an LED by varying the duty cycle of the on signal to the LED.

Program 5.3 **Turns on an LED and dims the one next to it.**

```
CLEAR                           ; always start with a CLEAR
                                ; statement
DEFINE OSC 4                    ; osc speed
TRISD = %11111100               ; set only PORTD pin 0 and 1
                                ; to outputs
X VAR BYTE                      ; declare x as a variable
PORTD.1 = 1                     ; turned on LED1 to compare it
                                ; to LED0
                                ;
LOOP:                           ; start of loop
  FOR X = 1 TO 255 STEP 2       ; set up loop for x
    PWM PORTD.0, X, 3           ; vary the duty cycle
    PAUSE 200/X                 ; pauses longer for the dimmer
                                ; values.
  NEXT X                        ; end of loop for x
GOTO LOOP                       ; return and do it again
END                             ; all programs must end with an
                                ; END statement
```

With the preceding programs we learned that we can control the on-off state and the brightness on an LED. Controlling the brightness becomes relevant when we are controlling seven segment displays because the LEDs in them are turned on one at a time and the duty cycle has to be managed properly to get an acceptable display within an acceptable time frame.

The LCD Display

This section describes the use of and interactions with existing hardware connections as they come with the LAB-X1 module. Other wiring schemes can be used with ease as defined in the compiler manual.

The LCD is controlled from PORTD, and all eight bits of this port are connected to the LCD. You therefore have the choice of using only the four high bits as a 4-bit data path for the LCD or using all eight bits. The entire port is also connected to eight of the LEDs on the 10-light LED bar graph. (The two leftmost LEDs in the bar graph are used to indicate that the power to the LAB-X1 is on.) The four high bits, bits D4 to D7, cannot be used for any other purpose if the LCD is being used. The software does not release these four bits automatically after using them to transfer information to the LCD, but you do have the option of saving the value of PORTD before using the LCD and then restoring this value after the LCD has been written to. The complication, of course, is that there will be a short glitch when the LCD is written to, and the use you make of PORTD has to tolerate this discontinuity.

PORTE, which has only three external lines, is dedicated to controlling the information transfer to the LCD. These lines can be used for other purposes (analog or digital) if the LCD is not being used.

The LCD provided on the LAB-X1 allows us to display two lines of 20 characters each. Its connections to the microcontroller are shown on the schematic provided with the LAB-X1 and are as shown in Figure 5.3.

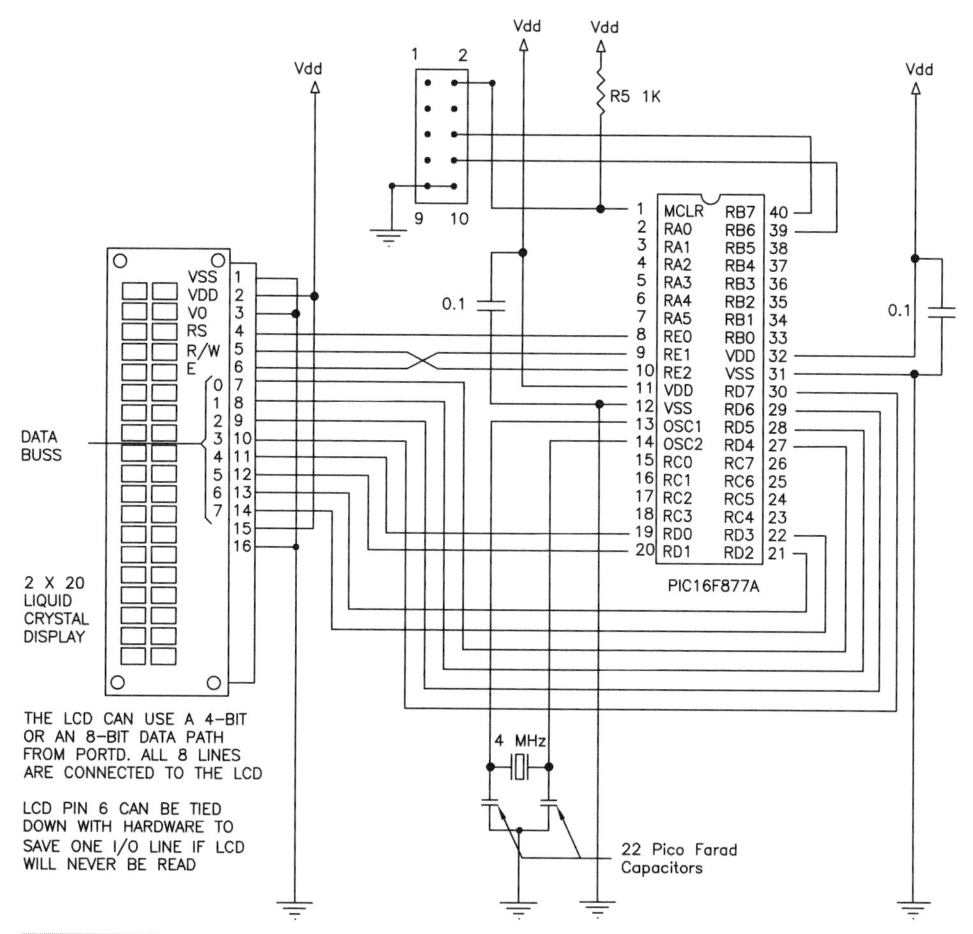

Figure 5.3 The LCD display wiring

Figure 5.3 is an easy to comprehend schematic diagram that shows the lines between the microcontroller and the display module. The other wiring is still in place, but it has been suppressed so we can concentrate on the LCD connections.

In Figure 5.3 we see that the LCD uses all the lines available on ports D and E. All of PORTD is used as the port the data will be put on, and PORTE, which has only three lines, is used to control data transfer to the LCD. We also know from looking at the full schematics provided with the LAB-X1 that all of PORTD is also connected to the LED bar graph. This does not affect the programming of the LCD and we will ignore this for now. You will, however, notice that the LEDs in the bar graph go on and off as programs run because we will be manipulating the data on these lines (D0 to D7). It is also possible to control the LCD with just the four high bits of PORTD, and we will use the scheme for most of the programs in this book. See the PBP manual for more information on how this is done.

Let us write the ubiquitous "Hello World" program for the LCD as our first exercise in programming the LCD. Once we know how to do that, we can basically write whatever we want to the LCD display and whenever we want to.

Before we can write to the LCD we have to define how the LCD is connected to the MCU. Also, since the 16F877A has some analog capabilities, it always starts up and resets in its analog mode, and it has to be put into digital mode for (at least) PORTE before we can use any of the digital properties of the affected ports (A and E).

The compiler manual says that we have to specify the location of both the LCD data and the LCD control lines that connect it to the system so that the compiler can address the device properly. Doing so allows us to place the LCD where convenient for us, in memory (the I/O lines), when we design our own devices, and the compiler will be able to address the LCD. The ports and lines used are specified in DEFINE statements that must be executed early in the program before the LCD is addressed.

Note *When you are designing your own devices it will be an advantage to place your LCD at the same memory locations used by the LAB-X1 so that your programs will run on the LAB-X1 for testing purposes, should you get into trouble. Being able to run the program on the LAB-X1 will let you know if it is a hardware or a software problem. All the devices I built used the same addresses as the LAB-X1 for the LCD, and this is reflected in the programs listings throughout this book.*

For the LCD display registers, on the LAB-X1, the DEFINE statements are as indicated in Program 5.4.

Program 5.4 **Displaying and blinking "HELLO WORLD" in the LCD display**

```
CLEAR                         ; define LCD registers and
                              ; control bits
DEFINE OSC 4                  ; osc speed
DEFINE LCD_DREG PORTD         ; data register
```

(continued)

Program 5.4 **Displaying and blinking "HELLO WORLD" in the LCD display** (*continued*)

```
DEFINE LCD_RSREG PORTE      ; select register          ]
DEFINE LCD_RSBIT 0          ; select bit               ] These
                                                          defines
DEFINE LCD_EREG PORTE       ; enable register          ] are all
                                                          explained
DEFINE LCD_EBIT 1           ; enable bit               ] in the PBP
DEFINE LCD_RWREG PORTE      ; read/write register      ] manual
DEFINE LCD_RWBIT 2          ; read/write bit           ]
DEFINE LCD_BITS 8           ; width of data path       ] can also
                                                          use 4
DEFINE LCD_LINES 2          ; lines in display         ]
DEFINE LCD_COMMANDUS 2000 ; delay in micro seconds ]
DEFINE LCD_DATAUS 50        ; delay in micro seconds ]
PAUSE 500                   ; to allow the LCD to initialize
; Set the port directions. We are setting (must set) all of
; PORTD and all of PORTE as outputs even though PORTE has
; only 3 lines. The other 5 lines will be ignored by the
; system.
                            ;
TRISD = %00000000           ; set all PORTD lines to output
TRISE = %00000000           ; set all PORTE lines to output
                            ; set the Analog to Digital control
                            ; register
ADCON1 = %00000111          ; needed for the 16F877A see notes
                            ; on pages 50 and 52.
                            ; this makes all of ports A and E
                            ; digital.
LOOP:                       ; the main loop of the program
  LCDOUT $FE, 1             ; clear screen, go to position 1
                            ; line 1
  PAUSE 250                 ; pause 0.25 seconds
  LCDOUT "HELLO"            ; print
  LCDOUT $FE, $C0           ; go to second line, 1st position
  LCDOUT "WORLD"            ; print
  PAUSE 250                 ; pause 0.25 seconds to see the
                            ; display
GOTO LOOP                   ; repeat
END                         ; all programs must end in END
```

Program 5.4 demonstrates the most elementary control over output to the LCD display. Variations of these lines of code will be used to write to the LCD in all our programs. (We will always use these addresses but when the reader writes his or her programs they can be at any suitable address.) Be sure to include the PAUSE 500 instruction in all your programs to allow the LCD enough time to initialize.

Not all the preceding DEFINE statements are needed on the LAB-X1, and you will notice this in some of the sample programs in this book, but when you build your own devices, you will need to include them all to make sure that nothing has been omitted.

```
ADCON1 = Analog to Digital CONtrol register #1.
```

The ADCON1=%00000111 statement, or one like it, is needed for our use of the 16F877A because any PIC MCU processor that has any analog capabilities comes up in the analog mode on reset and startup. In the analog mode all the lines of the PIC that have analog capabilities are set to the analog mode. This particular instruction puts all the analog pins on ports A and E into the digital mode. Since we need only PORTE and PORTD for controlling the LCD, none of PORTA needs to be in digital mode. I am showing %00000111 because all the examples provided by MicroEngineering Labs use this value. See the data sheet for more detailed information. (The use of this register is explained in Table 9-8 on using the LCD.) If you want to turn just the three available lines on PORTE to digital mode, you can use any binary value from 010 to 111 inclusive.

The control of the A to D conversion capability is managed by the four low bits of ADCON1. For our purposes, bit 0 and bit 3 are not relevant.

Note *Much of the information in this chapter can be found on page 126 in Section 11 Analog to Digital Converter (A/D) Module of the data sheet.*

The following lists how the four least significant bits in register ADCON1 are used to manage the A to D setting of the three bits of PORTE and five relevant bits of PORTA. (We are setting them all except PORTA.4 to digital.) Bit 0 is not relevant to the LCD operation (it is a "don't care" bit).

- Bit 1 and 2 must be set to 1 to make the two ports (A and E) digital
- Bit 3 is not relevant to the LCD operation (it too is a "don't care" bit.)

So ADCON1 = %00000110 or %00000111 would be adequate for our work. (We could also have done this in decimal format with ADCON1=6 or with ADCON1=7.)

Writing Binary, Hex, and Decimal Values to the LCD

The value of numbers written to the LCD can be specified with the following prefixes that determine whether the value will be displayed as a binary, a hexadecimal, or a decimal value and to specify how many digits will be displayed. See the PBP manual for details.

- BIN specifies that the display will be binary
- HEX specifies that the display will be in hexadecimal format
- DEC specifies display in decimal format.

In Program 5.5, the value of NUMB is set to 170 because it alternates 1s and 0s in binary format. Any number below or equal to 255 could have been used. Using BIN8 instead of BIN displays all eight bits. Using HEX2 instead of HEX displays both hex digits. DEC5 can display all five decimal digits because we are limited to 16 bits (65535) and integer math in PICBASIC. BIN16 can be used for 2-byte words to display all 16 bits. As previously stated, any number of digits can be displayed. Program 5.5 demonstrates the possibilities. (Note that we are not looping around the display instruction in this program.)

Program 5.5 **Writing to the LCD display in FULL binary, hexadecimal, and decimal**

```
CLEAR                              ; clear memory
DEFINE OSC 4                       ; osc speed
DEFINE LCD_DREG PORTD              ; define LCD connections
DEFINE LCD_DBIT 4                  ; define LCD connections
DEFINE LCD_RSREG PORTE             ; define LCD connections
DEFINE LCD_RSBIT 0                 ; define LCD connections
DEFINE LCD_EREG PORTE              ; define LCD connections
DEFINE LCD_EBIT 1                  ; define LCD connections
ADCON1 = %00000110                 ; Make PORTA and PORTE digital
LOW PORTE.2                        ; LCD R/W low (write) we will
                                   ; do no reading
PAUSE 500                          ; wait for LCD to start
                                   ;
NUMB VAR BYTE                      ; assign variable
                                   ;
TRISD =  %00000000                 ; D7- -D0 area all outputs
NUMB = %10101010                   ; this is decimal 170
                                   ;
LCDOUT $FE, 1                      ; clear the LCD
LCDOUT $FE, $80, BIN8 NUMB," ",HEX NUMB, " ", DEC5 NUMB," "
                                   ; display numbers
END                                ; end program
```

READING A POTENTIOMETER AND DISPLAYING THE RESULTS ON THE LED BAR GRAPH

On the PIC 16F877A each potentiometer is placed across five volts and ground. (Other reference voltages and resistances can also be used. See the data sheet.)

When we read a potentiometer, the MCU divides the voltage across the potentiometer into 256 steps between 0 and 255 and gives us the number that represents the position of the wiper across the connected voltage. Neither the voltage nor the resistance of the potentiometer is relevant (though it can be if we know the minimum and maximum voltage across the pot). What we are getting is the relative position of the wiper expressed as an 8-bit number. (The PIC also has 10-bit resolution capability; see the data sheet.)

On the LAB-X1 each of the three potentiometers is placed across five volts and ground, and their wipers are connected to three PORTA lines. (The circuitry for this is shown in Figure 5.4). The potentiometers are read as 8-bit values with a built-in 8-bit A to D converter. This gives a full scale reading of 0 to 255 for each of the three potentio meters no matter what the actual total resistance value of the potentiometer. If you want to read the resistance in ohms, you have to divide the reading by 255 and multiply by the total resistance of the potentiometer. (The potentiometer value has to be high enough so that the potentiometer does not act as a short between ground and the MCU power connection. 5K to 10K ohms is okay for most purposes.) If extremely high resistances are used, the readings can become jittery.

In Program 5.6 we will read one of the potentiometers (the one nearest the edge of the board) to an accuracy of 8 bits and display the results on the rightmost 8 LEDs of the LED bar graph. This potentiometer is connected to pin 2 of the PIC (also identified as RA0 and as pin PORTA.0). We will display the result of the value read (0 to 255) on the bar graph by loading the reading into PORTD. Since PORTD is connected to the eight LEDs, this will automatically give us a binary reading of the data. In the next step, we will display the information on the LCD display as alphanumeric data (which is of course much easier to read)

We can expand the program to not only display on the bar graph, but also put the information on the LCD display. However, there's a problem in that the PORTD lines are shared by the bar graph and the LCD display. When we run the program we will notice there is a background noisy blinking of the LEDs in the bar graph as the LCD is being written to, but after that finishes the bar graph displays the data from the potentiometer as expected. If we had hardware and software that could suppress the LEDs when we were writing to the LCD, this problem could be eliminated. The operation observed demonstrates that the chip select line allows us to use the lines of PORTD to control both the LED bar graph and the LCD display. Notice that the delay (that allows us to read the display) has to come immediately after setting PORTD to A2D_Value for this to work properly. When we do each step, where we put the pauses is important when using microprocessors. Program 5.6 implements the preceding procedure.

Program 5.6 **Displaying the potentiometer wiper position on the LCD and the LED bar graph**

```
CLEAR                      ; define LCD connections
DEFINE OSC 4               ; osc speed
DEFINE LCD_DREG PORTD      ; define LCD connections
DEFINE LCD_DBIT 4          ; define LCD connections
DEFINE LCD_RSREG PORTE     ; define LCD connections
DEFINE LCD_RSBIT 0         ; define LCD connections
DEFINE LCD_EREG PORTE      ; define LCD connections
DEFINE LCD_EBIT 1          ; define LCD connections
ADCON1 = %00000110         ; make PORTA and PORTE digital
LOW PORTE.2                ; LCD R/W low (set it to write only)
PAUSE 500                  ; wait for LCD to start up
                           ;
```

(*continued*)

Program 5.6 **Displaying the potentiometer wiper position on the LCD and the LED bar graph** (*continued*)

```
NUMB VAR BYTE              ; assign variable
                           ;
TRISD = %00000000          ; D7 to D0 outputs
A2D_VALUE VAR BYTE         ; create A2D_Value to store result
TRISA = %11111111          ; set PORTA to all input
ADCON1 = %00000010         ; set PORTA to analog input
                           ;
LCDOUT $FE, 1              ; clear the LCD
                           ; define the ADCIN parameters
DEFINE ADC_BITS 8          ; set number of bits in result
DEFINE ADC_CLOCK 3         ; set clock source (3=rc)
DEFINE ADC_SAMPLEUS 50     ; set sampling time in µS
                           ;
LOOP:                      ; start loop
  ADCIN 0, A2D_VALUE       ; read channel 0 to A2D_Value
  LCDOUT $FE, $80, "VALUE = ", HEX2 A2D_VALUE, " ", DEC5
 A2D_VALUE
  LCDOUT $FE, $C0, BIN8 A2D_VALUE   ;
  PORTD = A2D_VALUE        ; the pause must come right after
                             setting
  PAUSE 250                ; PORTD and before PORTD is used again
                           ; try setting PORTD before the LCDOUT
GOTO LOOP                  ; do it forever
END                        ; end program
```

The information read from potentiometer 0 is displayed on the bar graph in Program 5.6. The other two potentiometers are being ignored. Note that pin 4 was skipped in the circuitry shown in Figure 5.4. This pin does not have an analog capability.

SIMPLE BEEP

We have one other piece of hardware that we can output to, and that is the small piezo electric speaker on the board. This speaker is connected to line PORTC.2.

The PWM (pulse width modulation) command can be used to create a short beep on the piezo electric speaker on the LAB-X1. The command specifies the PORTC pin to be used, the duty cycle and the duration of the beep (100 milliseconds in this case). Program 5.7 demonstrates how the speaker is used.

Program 5.7 **Generate a short tone on the piezo speaker.**

```
CLEAR                        ; clear memory
DEFINE OSC 4                 ; osc speed
PWM PORTC.2, 127, 100        ; beep command
END                          ; end the program
```

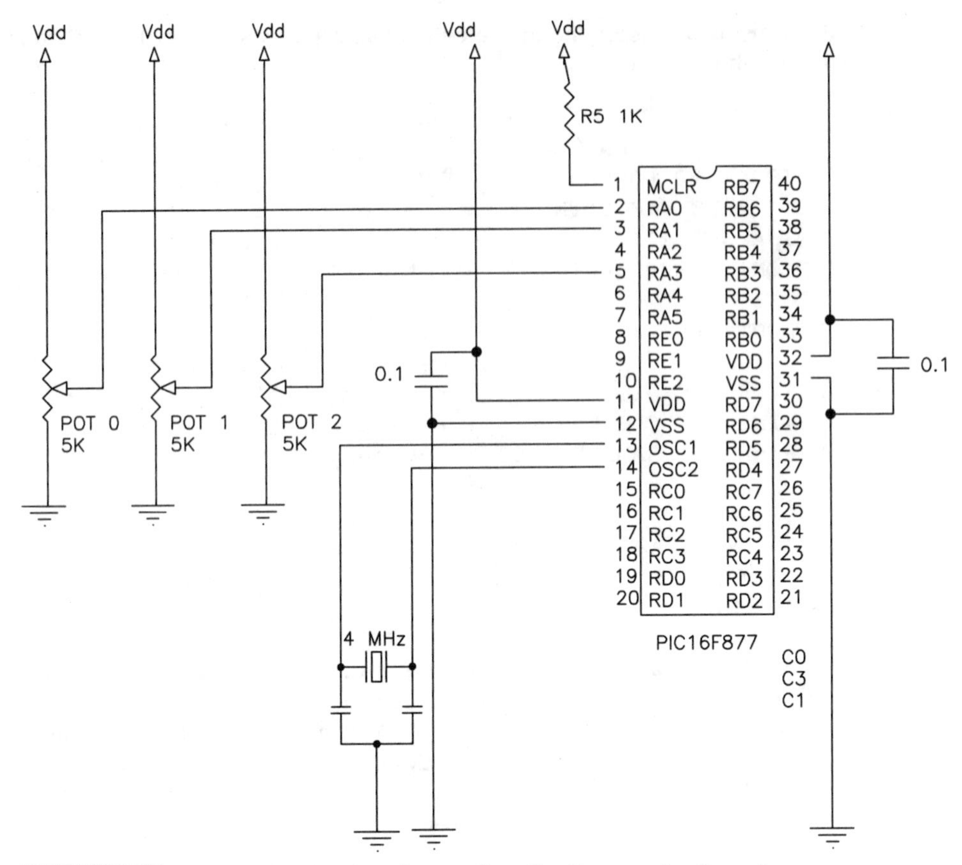

Figure 5.4 The basic circuitry for reading the three potentiometers

Program 5.7 provides a 0.1 second (100 milliseconds) beep. Press the reset button to repeat the beep.

Check to see what happens if you leave the END statement off in Program 5.7.

Be sure to note the following about Program 5.7:

- Program 5.7 generates a 50 percent duty cycle for 100 cycles.
- It defines that you are using a 4 MHz oscillator.
- PORTC.2 specifies the pin to be used.
- 127 specifies a 50 percent duty cycle; the range of the variable is from 0 to 255.
- 100 specifies that the tone is to last for a 100 each of the 256 on-off steps that define one cycle.

In the PWM command the frequency and length of the signal generated are dependent on the oscillator frequency. In this case this is 4 MHz, and one cycle is about 2.5 millisecond long (0.0025 seconds).

Note that the line C2 is also connected to the output for a possible phone jack and to an IR LED that can interact with IR receivers. These two connections are not populated on the PC board as received, but they can be added with little difficulty.

There are two types of signals that can be annunciated on the speaker as programmed from the compiler. The PWM command can send a signal of a fixed duty cycle for a fixed number of cycles, and the HPWM (hardware PWM) command can set up a PWM signal that runs continuously in the background. In either case, the signal needs to be provided on the PORTC.2 pin because that is where the speaker is connected. However, the normal PWM (not the background HPWM) command signal can be made to appear at any available pin. The background HPWM signal can be modified "on the fly" in a program. We will use this feature to modulate the power to the motors, under program control, when we start running motors.

The HPWM signals can only be made available at pin PORTC.2 (Channel 1) and PORTC.1 (Channel 2). Yes, the pin numbers are reversed! In the PIC 16F877A there are only two HPWM channels and these two pins are connected permanently to these two channels. (Some PIC devices provide more than two channels. See the data sheets.) Since we have the speaker hard wired to PORTC.2, we can use only Channel 1 for the tones we generate.

As seen in Figure 5.5, these signals can also be used to generate telephone dial tones (DTMF) and infrared (IR) signals when provided with the appropriate hardware.

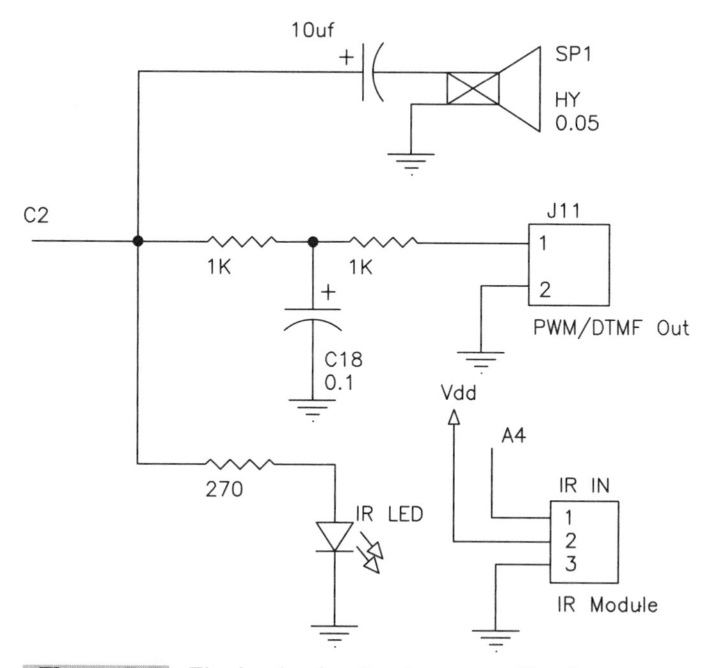

Figure 5.5 **The basic circuitry for generating tones on the piezo speaker on the hardware provided. If you use an infrared receiver its signal will appear on line A4 as shown above. Only relevant circuitry is shown.**

We will concentrate on creating tones on the piezo speaker. The wiring and programming is the same for the other devices. All we have to do is change the parameters.

A slightly more complicated program, Program 5.8 demonstrates the use of PWM to control the brightness of one of the LEDs in the bar graph.

Program 5.8 **LED dimming using the PWM command**

```
CLEAR                              ; clear memory
DEFINE OSC 4                       ; osc speed
TRISD = %11111110                  ; set only PORTD pin 1 to
                                   ; output
X VAR BYTE                         ; declare x as a variable
                                   ;
LOOP:                              ; start loop
  FOR X = 0 TO 255 STEP 5          ; ] in this loop the value
    PWM PORTD.0, X, 3              ; ] x represents the brightness
  NEXT X                           ; ] of the LED at PORTD.0
GOTO LOOP                          ; repeat loop
END                                ; end program
```

Using the HPWM command is a bit more complicated in that you have to define certain parameters before we can use the command. The necessary defines are as follows:

```
DEFINE CCP1_REG PORTC              ; port to be used by HPWM 1
DEFINE CCP1_BIT 2                  ; pin to be used by HPWM 1
DEFINE CCP2_REG PORTC              ; port to be used by HPWM 2
DEFINE CCP2_BIT 1                  ; pin to be used by HPWM 2
```

You also have to define which timer the signal will use so that other timers can be used for other purposes while the signal is being generated. If a timer is not specified the system defaults to Timer1, the 16-bit timer.

The command is as follows:

```
HPWM Channel, DutyCycle, Frequency
```

The following commands create a 50 percent duty cycle PWM signal at 1500 Hz (as affected by the definition of OSC) on PORTC.2 continuously in the background:

```
DEFINE OSC 4               ; osc speed
HPWM 1, 127, 1500          ; generate background PWM.
```

See Program 5.9 for a complete listing that demonstrates the use of these instructions in a real situation.

The command can be updated in run time from within the program. As might be expected, the pin cannot be used for any other purpose as long as it is generating the PWM signal. Turn off the PWM mode at the CCP control register to use the pin as a normal pin. See the data sheet for more information.

The frequencies generated are limited by the frequency of the oscillator being used to clock the PIC processor. The minimum frequency for the PIC 16F877A is 1221 Hz (with a 20 MHz oscillator). See the PICBASIC PRO Compiler manual for more information on other frequencies.

Program 5.9 **Generates a tone on the piezo speaker**

```
CLEAR                       ; clears memory
DEFINE OSC 4                ; osc speed
DEFINE CCP1_REG PORTC       ; port to be used by HPWM 1
DEFINE CCP1_BIT 2           ; pin to be used by HPWM 1
                            ; since no timer is defined, Timer1
                              will be used;
HPWM 1,127,2500             ; the tone command
PAUSE 100                   ; pause 0.1 second to hear tone
END                         ; end program to stop tone.
```

Next, in Program 5.10, we will generate some telephone touch tones on the speaker to demonstrate the capability provided by the DTMFOUT command:

```
DTMFOUT Pin, {Onms, Offms} [Tone#{Tone#...}]
```

Since we will be using pin C2, our command will look similar to the above because we are using default values for ONms and OFFms

Program 5.10 **Generates telephone key tones on the piezo speaker (555-1212)**

```
CLEAR                                    ; clear memory
DEFINE OSC 4                             ; osc speed
DTMFOUT PORTC.2, [5, 5, 5, 1, 2, 1, 2]   ; telephone tones
END                                      ; end program
```

The key tones generated are rough (before filtering) but you can tell that they mimic the telephone dialing tones. The signal needs to go through a filter and then an amplifier to be clean and viable. There are a number of constraints on this use of this command depending on the processor being used and the speed of the oscillator in the circuit. See the PICBASIC PRO Compiler manual for details.

The FREQOUT command can also be used to generate telephone dialing frequencies. See the PICBASIC PRO Compiler manual for details.

CONTROLLING AN RC SERVO FROM THE KEYBOARD

Now that you know how to generate pulses and read the potentiometers, you can use the LAB-X1 to control the position of an RC servo connected to port J7 from switches SW1, 2, and 3 on the keyboard. The program is to be designed such that:

- Switch 1 will turn the servo clockwise incrementally.
- Switch 2 will center the servo.
- Switch 3 will turn the servo counterclockwise incrementally.

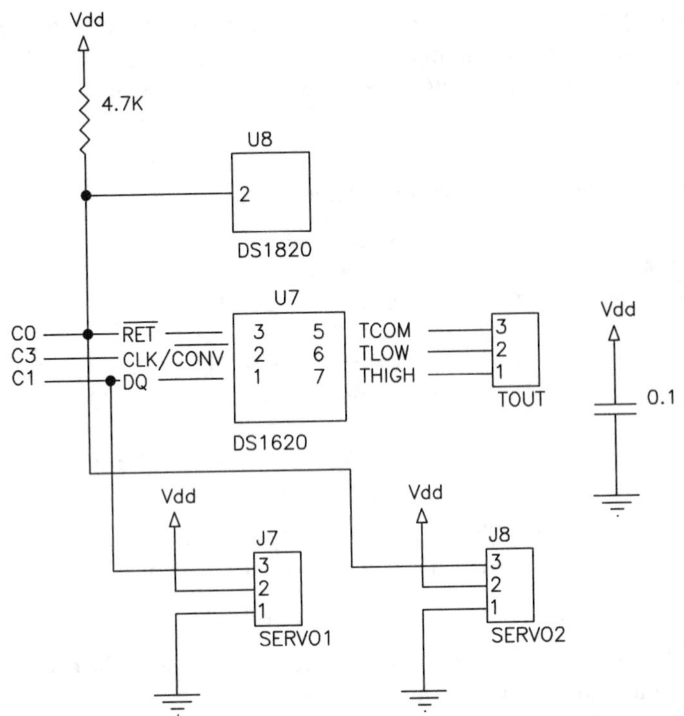

Figure 5.6 **Circuitry for controlling an RC servo from the three potentiometers**

Note that by changing a few variables that are defined up front in Program 5.11, you can adjust the center position, the incremental step value, and the extreme CW and CCW positions of the servo. (This program has been adapted from, and made simpler than, a program in the MicroEngineering Labs sample programs. It is instructive to compare this program with programs SERVOX and SERVO1in the sample programs.) The circuitry that controls that can be used to control the two ports that the R/C servos can be connected to is shown in Figure 5.6.

As always, only relevant components are shown in Figure 5.6. Connect the servos to jumper J7. The circuitry shown in Figure 5.6 is used in Program 5.11.

Program 5.11 **Servo position control for an R/C servo from PORTB buttons.** This program uses a servo at Jumper J7.

```
CLEAR                                   ; clear memory
DEFINE OSC 4                            ; osc speed
DEFINE LCD_DREG PORTD                   ; define LCD connections
DEFINE LCD_DBIT 4                       ;
DEFINE LCD_RSREG PORTE                  ;
DEFINE LCD_RSBIT 0                      ;
DEFINE LCD_EREG PORTE                   ;
```

(continued)

Program 5.11 Servo position control for an R/C servo from PORTB buttons. This program uses a servo at Jumper J7 (*continued*).

```
DEFINE LCD_EBIT 1                    ;
POS VAR WORD                         ; servo position variable
CENTERPOS VAR WORD                   ; servo position variable
MAXPOS VAR WORD                      ; servo position variable
MINPOS VAR WORD                      ; servo position variable
POSSTEP    VAR BYTE                  ; servo position step
                                     ; variable
SERVO1     VAR PORTC.1               ; alias servo pin Use J7
                                     ; for servo
POS = 0                              ; set variables
CENTERPOS = 1540                     ; set variables
MAXPOS = 2340                        ; set variables
MINPOS = 740                         ; set variables
POSSTEP = 5                          ; set variables
ADCON1 = %00000111                   ; PORTA and PORTE to
                                     ; digital
LOW PORTE.2                          ; LCD R/W low = write
PAUSE 100                            ; wait for LCD to startup
OPTION_REG = $01111111               ; enable PORTB pullups
LOW SERVO1                           ; servo output low
GOSUB CENTER                         ; center servo
LCDOUT $FE, 1                        ; clears screen only
                                     ;
MAINLOOP:                            ; main program loop
  PORTB = 0                          ; PORTB lines low to
                                     ; read buttons
  TRISB = $11111110                  ; enable first button row
  IF PORTB.4 = 0 THEN GOSUB LEFT     ; check if any button
                                     ; pressed to move servo
  IF PORTB.5 = 0 THEN GOSUB CENTER   ;
  IF PORTB.6 = 0 THEN GOSUB RIGHT    ;
  LCDOUT $FE, $80, "POSITION = ", DEC POS , " " ;
  SERVO1 = 1                         ; start servo pulse
  PAUSEUS POS                        ;
  SERVO1 = 0                         ; end servo pulse
  PAUSE 16                           ; servo update rate
                                     ; about 60 Hz
GOTO MAINLOOP                        ; do it all forever
                                     ;
LEFT:                                ; move servo left
  IF POS < MAXPOS THEN POS = POS + POSSTEP      ;
RETURN                               ;
                                     ;
RIGHT:                               ; move servo right
  IF POS > MINPOS THEN POS = POS - POSSTEP      ;
```

(continued)

Program 5.11 Servo position control for an R/C servo from PORTB buttons. This program uses a servo at Jumper J7 (*continued*)

```
RETURN                                    ;
                                          ;
CENTER:                                   ; center servo
POS = CENTERPOS                           ;
RETURN                                    ;
END                                       ; end program
```

Now we will make Program 5.11 more sophisticated by using the three potentiometers on the LAB-X1 to manipulate the three variables that control the center position, the end positions, and the incremental move of the servo in the preceding program. We will use just one variable to adjust both the end positions because we have only three potentiometers. If we had four potentiometers we could make the adjustment to the limits on one side independent of the adjustment to the other.

The control functions we will implement are described next:

- We will allow the center position to be adjusted by 127 counts in each direction.
- The end positions will be made variable by 127 counts at each end.
- The incremental move will be adjustable from 1 to 20 counts per keypress.

First we will make it possible to read the potentiometers. We already know how to do this. Then we will add the math relationships to the variables in the program so that the readings from the potentiometers interact with the three variables appropriately.

The three potentiometers will be assigned as described next.

POT1, the one nearest the board edge, controls the center position.

POT2, the central pot, controls the limit positions.

POT3 sets the speed of the servo by setting the step amount.

Program 5.12 Uses an R/C servo connected to jumper J7 Servo position control, with added functions

```
CLEAR                          ; clear memory
DEFINE OSC 4                   ; osc speed
DEFINE LCD_DREG PORTD          ; define LCD connections
DEFINE LCD_DBIT 4              ;
DEFINE LCD_RSREG PORTE         ;
DEFINE LCD_RSBIT 0             ;
DEFINE LCD_EREG PORTE          ;
DEFINE LCD_EBIT 1              ;
DEFINE ADC_BITS 8              ; set number of bits in result
DEFINE ADC_CLOCK 3             ; set clock source (3=rc)
DEFINE ADC_SAMPLEUS 50         ; set sampling time in µS
TRISA = %11111111              ; set PORTA to all input
TRISD = %00000000              ; set all PORTD lines to outputs
```

(continued)

Program 5.12 **Uses an R/C servo connected to jumper J7 Servo position control, with added functions** *(continued)*

```
ADCON1 = %00000111              ; PORTA and PORTE to digital
LOW PORTE.2                     ; LCD R/W line low (W)
A2D_VALUE  VAR BYTE             ; create A2D_Value to store
                                ; result
A2D_VALUE1  VAR BYTE            ; create A2D_Value to store
                                ; result
A2D_VALUE2  VAR BYTE            ; create A2D_Value to store
                                ; result
ADWALWAS  VAR BYTE              ; remembers A/D value
POS  VAR  WORD                  ; servo positions
CENTERPOS  VAR WORD             ; center position
MAXPOS  VAR WORD                ; max position
MINPOS  VAR WORD                ; min position
POSSTEP  VAR BYTE               ; position step
PAUSE 500                       ; wait .5 second
SERVO1 VAR PORTC.1              ; alias servo pin
ADCIN 0, A2D_VALUE              ; read channel 0 to A2D_Value
OPTION_REG = $7F                ; enable PORTB pull ups
LOW SERVO1                      ; servo output low
GOSUB CENTER                    ; center servo
LCDOUT $FE, 1                   ; clears screen only
PORTB = 0                       ; PORTB lines low to read
                                ; buttons
TRISB = %11111110               ; enable first button row
                                ;
MAINLOOP:                       ; main program loop
                                ; check any button pressed to
                                ; move servo
  IF PORTB.4 = 0 THEN GOSUB LEFT    ;
  IF PORTB.5 = 0 THEN GOSUB CENTER  ;
  IF PORTB.6 = 0 THEN GOSUB RIGHT   ;
  ADCIN 0, A2D_VALUE    ; read channel 0 to A2D_Value
  ADCIN 1, A2D_VALUE1   ; read channel 1 to A2D_Value 1
  ADCIN 3, A2D_VALUE2   ; read channel 2 to A2D_Value 2
  MAXPOS = 2350 -127 + A2D_VALUE1   ;
  MINPOS = 750 +127-A2D_VALUE1      ;
  CENTERPOS = POS-127 + A2D_VALUE   ;
  POSSTEP = A2D_VALUE2/13 +1    ;
  SERVO1 = 1                    ; start servo pulse
    PAUSEUS POS                 ;
  SERVO1 = 0                    ; end servo pulse
  LCDOUT $FE, $80, "POS=", DEC POS-127 + A2D_VALUE , " ",
DEC A2D_VALUE," ",_DEC A2D_VALUE1," " ,DEC POSSTEP," ";
  PAUSE 16                      ; servo update rate about 60 Hz
```

(continued)

Program 5.12 **Uses an R/C servo connected to jumper J7 Servo position control, with added functions** (*continued*)

```
GOTO MAINLOOP                          ; do it all forever
                                       ;
LEFT:                                  ; move servo left
  IF POS < MAXPOS THEN POS = POS + POSSTEP
RETURN                                 ;
                                       ;
RIGHT:                                 ; move servo right
  IF POS > MINPOS THEN POS = POS - POSSTEP
RETURN                                 ;
                                       ;
CENTER:                                ; center servo
  POS = 1540-127 + A2D_VALUE           ;
RETURN                                 ;
                                       ;
END                                    ; end program
```

At this stage we are starting to get an idea about how one might take a simple problem and make it more amenable to a more sophisticated solution by adding simple hardware and software features to it. In Program 5.12 we have gone from a simple but rigid control of the position of a servo to a much more flexible and user friendly approach.

READING THE INPUTS

Now that we are beginning to learn how to control the output, we need to learn how to read the inputs and manipulate the outputs based on what the input was. In other words, we are going to learn how to create interactive, and thus maybe more useful programs.

In the first input program we will read the first column, first row pushbutton (SW1) and turn on an LED only while the button (SW1) is down.

The simplest input is one pushbutton and the simplest output is one LED turned on. We will use just these two devices but we will add a little complication. The LED is to be programmed to be on only while the button is held down. We will use button 1 (top left) on the keyboard and the LED connected to PORTD.0. But first we need to learn how to read a switch on the keyboard.

Reading the Keyboard

On the LAB-X1, PORTB is dedicated to the interface with the keyboard. Lines B0 to B3 are connected to the rows, and lines B4 to B7 are connected to the columns of the keyboard. When the keyboard is not being used, the lines may be used for other purposes, keeping in mind the internal pull up capability and the in-line load limiting resistors on the lower four bits/lines (B0 to B3) (which can easily remain outside our circuitry when we design circuitry for other purposes).

The keyboard is connected to PORTB such that the columns of the keyboard matrix are connected to the high nibble of a port and the rows are connected to the low nibble. The wiring is shown in Figure 5.7.

To read the keyboard, the low nibble of PORTB is set to be outputs and the high nibble is set to be inputs.

PORTB has a special property that allows its lines to be pulled high (very weakly) with internal resistors by setting OPTION_REG.7 = 0. This property of the port can affect all the bits (B0 to B7), but only those bits that are actually programmed to be inputs with TRISB will be affected.

Note *Pins B4 to B7 can be programmed to generate interrupts and B0 can be programmed to awaken a sleeping PIC.*

Next, the four lines (the low bits B3 to B0) are made low one at a time and the high bits are polled to see if any of them has been pulled low. If any of the switches is (because only one can be read at one time) down, one of the lines will be pulled low. Because we know which low nibble bit was low when the high nibble bit became low we can determine which key has been pressed. For our purposes, at this stage, we are interested only in SW1, the upper left switch, so we can simplify the diagram to what we see in Figure 5.8.

In Figure 5.8, we can see that if we make PORTB.0 low and PORTB.4 had been pulled high, PORTB.4 will become low if SW1 is held down. No polling is necessary at this stage. Once the conditions are set up, all we have to do is create a loop that turns the PORTD.0 LED on if the switch SW1 is down and off for all other conditions. Program 5.13 demonstrates one way of doing this.

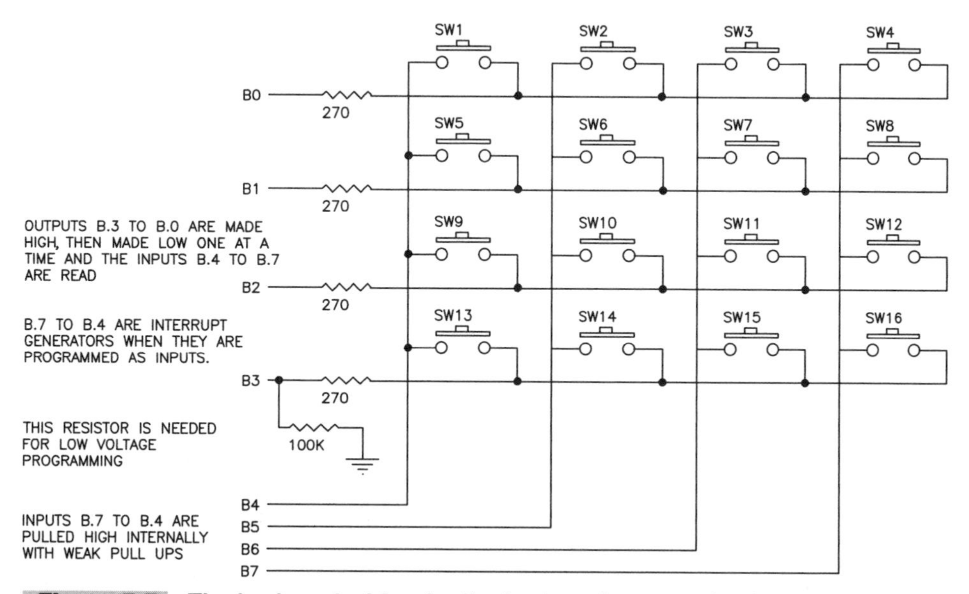

Figure 5.7 The keyboard wiring for the keyboard rows and columns

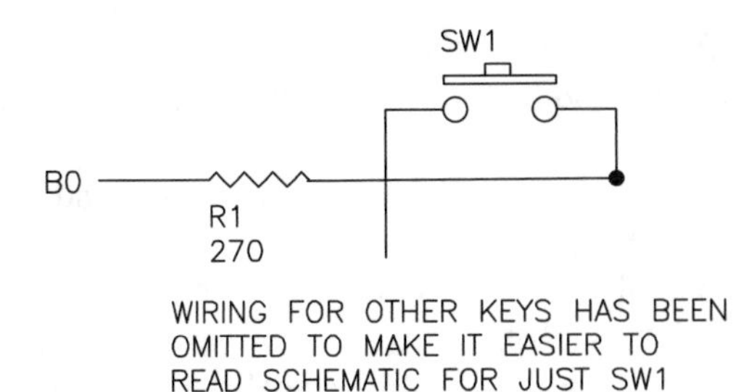

Figure 5.8 **Partial keyboard, the wiring for just one switch SW1. The other wiring is still there but is being ignored in the diagram and in the program.**

Program 5.13 **Reading a switch.** Read SW1 and turn LED on PORTD.0 on while it is down.

```
CLEAR                        ; clear memory
DEFINE OSC 4                 ; osc speed
TRISB = %11110000            ; set the PORTB directions
PORTB = %11111110            ; set only B0 made low.
                             ; see page 31 of the data sheet
                               for setting the pull ups
                             ; on PORTB
OPTION_REG.7 = 0             ; bit 7 of the OPTION_REG sets
                               the pull ups
                             ; when cleared
TRISD = %11111110            ; set only PORTD.0 to an output.
PORTD.0 = 0                  ; initialize this LED to off
                             ;
MAINLOOP:                    ; loop
  IF PORTB.4 = 1 THEN        ; check for first column being
                               low
    PORTD.0 = 0              ; if it is low turn D0 off
  ELSE                       ; decision
    PORTD.0 = 1              ; if not turn it on
  ENDIF                      ; end of decision check
GOTO MAINLOOP                ; repeat.
END                          ; all programs end in END
```

Remember, here we are looking at SW1 only. The other switches in this column will not turn the LED on because they are all high and cannot change the state of PORTB.4 because it is already pulled high (and needs to go low if we are to read it as having changed its state).

Read entire keyboard and display the binary value of the row and column read on the LCD.

Next, we learn how to read the entire keyboard and tell which key is pressed by identifying the active row and column numbers. This is a modification of the single key program with the scanning of the nibbles in PORTB added to determine what happened and when it happened.

A loop scans the high nibble of PORTB, which is the output from the keyboard. When all four bits were pulled high this nibble will be read as HEXF. If it is HEXF, no keys are down and we can rescan the keys. If, however, a key has been pressed, the answer will be other than HEXF and can be interpreted as follows:

- If B4 is low the answer will be HEXE (15–1=14) 1110 Column 1
- If B5 is low the answer will be HEXD (15–2=13) 1101 Column 2
- If B6 is low the answer will be HEXB (15–4=11) 1011 Column 3
- If B7 is low the answer will be HEX7 (15–8=7) 0111 Column 4

To determine which row the key that was pressed is in, we have to know which bit in the low nibble was taken low by the scanning routine.

The table of values for the low nibble is as follows:

- If B0 is low the low nibble will be HEXE (15–1=14) 1110 Row 1
- If B1 is low the low nibble will be HEXD (15–2=13) 1101 Row 2
- If B2 is low the low nibble will be HEXB (15–4=11) 1011 Row 3
- If B3 is low the low nibble will be HEX7 (15–8=7) 0111 Row 4

Having the two pieces of preceding information lets us identify the key that was pressed. Almost all keyboards use a scanning scheme similar to this. Often a PIC like MCU is dedicated to reading the keyboard and interrupting the main processor if a keystroke is detected.

In Program 5.14 we will display the contents of the entire byte (PORT B) on the first line of the LCD so we can actually see what is happening in the register represented by PORTB as we scan the lines. Then, on line 2, we will show the low byte and the high byte separately so we can see what each keypress does. We have added a 1/20 second delay in the loop (so that we can see the scanned value), so we have to hold each key down for over 1/20 second for the scan to make sure the keypress will register.

Program 5.14 Read keyboard. Read the keyboard rows and columns

```
CLEAR                             ; clear the memory
DEFINE OSC 4                      ; osc speed
DEFINE LCD_DREG PORTD             ; LCD defines
DEFINE LCD_DBIT 4                 ;
DEFINE LCD_RSREG PORTE            ;
DEFINE LCD_RSBIT 0                ;
DEFINE LCD_EREG PORTE             ;
```

(continued)

Program 5.14 **Read keyboard.** Read the keyboard rows and columns (*continued*)

```
DEFINE LCD_EBIT 1              ;
ADCON1 = %00000111             ; make PORTA and PORTE digital
LOW PORTE.2                    ; LCD R/W low (write)
PAUSE 500                      ; wait for LCD to start up
                               ;
READING VAR BYTE               ; define the variables
ALPHA VAR BYTE                 ; used as a counter
BUFFER VAR BYTE                ; stores PORTB when needed
                               ; Set up port B pull ups
OPTION_REG.7 = 0               ; enable PORTB pull ups to make
                               ; B4-B7 high
TRISB = %11110000              ; make B7-B4 inputs, B3-B0
                               ; outputs
BUFFER = %11111111             ; no key has been pressed for
                               ; display
                               ; set up the initial LCD
                               ; readings
LCDOUT $FE, 1                  ; clear the LCD
LCDOUT $FE, $C0, "ROW=",BIN4 (BUFFER & $0F)," COL=", BIN4
BUFFER >>4
                               ;
LOOP:                          ;
  PORTB = %00001110            ; set line B0 low so we can
                                 read row 1 only
  FOR ALPHA = 1 TO 4    ; need to look at 4 rows
    LCDOUT $FE, $80, BIN8 PORTB," SCANVIEW B"    ; see bits
                                                   scanned
    IF (PORTB & $F0)<>$F0 THEN ; as soon as one of the bits
                                 ; in B4 to B7 changes we
                                 ; immediately have to
                                 ; store the value of PORTB
      BUFFER = PORTB             ; in a safe place.
      GOSUB SHOWKEYPRESS         ;
    ELSE                         ;
    ENDIF                        ;
      PAUSE 50                   ; this pause lets us see the
                                 ; scanning but it also means
                                 ; that you have to hold a key
                                 ; down for over 50 µsecs to
                                 ; have it register. This pause
                                 ; can be removed after you
                                 ; have seen the bits
                                 ; scanning on the LCD
```

(*continued*)

Program 5.14 Read keyboard. Read the keyboard rows and columns (*continued*)

```
      PORTB = PORTB <<1         ; move bits left one place for
                                ; next line low
      PORTB = PORTB + 1         ; put 1 back in LSBit, the
                                ; right bit
  NEXT ALPHA                    ;
GOTO LOOP                       ;
                                ;
SHOWKEYPRESS:                   ; display
  LCDOUT $FE, $C0, "ROW=", BIN4 (BUFFER & $0F)," COL=", BIN4
BUFFER >>4
RETURN                          ;
END                             ; end program
```

Read Keyboard and Display Key Number on the LCD

Now that we understand how this works, we have to turn the binary information we have gathered into a number from 1 to 16 and identify the keypress on the LCD.

> **Note** *The sample program to do this provided on the Internet by MicroEngineering Labs shows another way of doing this and is worth studying. Seeing how different programmers address the same problem can be very instructive.*

The switch number is the row number plus (the column number –1) * 4.

If we reverse all the bits in the PORTB byte, the nibbles will give us the positions of the rows and columns as the locations of the 1s in the two nibbles. Make sure you understand this before proceeding. Work it out on a piece of paper step by step.

The Show Keypress subroutine used in Program 5.14 has to be modified as shown in Program 5.15.

Program 5.15 Reading the keyboard. Read the keyboard rows and columns and show key number.

```
CLEAR                    ; clear memory
DEFINE OSC 4             ; osc speed
DEFINE LCD_DREG PORTD    ; define LCD connections
DEFINE LCD_DBIT 4        ;
DEFINE LCD_RSREG PORTE   ;
DEFINE LCD_RSBIT 0       ;
DEFINE LCD_EREG PORTE    ;
DEFINE LCD_EBIT 1        ;
ADCON1 = 7               ; make PORTA and PORTE digital
LOW PORTE.2              ; LCD R/W low (write)
PAUSE 200                ; wait for LCD to start
                         ; define the variables
BUFFER VAR BYTE          ; define the variables
```

(*continued*)

Program 5.15 Reading the keyboard. Read the keyboard rows and columns and show key number (*continued*).

```
ALPHA VAR BYTE            ; define the variables
COLUMN VAR BYTE           ; define the variables
ROW VAR BYTE              ; define the variables
SWITCH VAR BYTE           ; define the variables
                          ; set up port b pull-ups
OPTION_REG.7 = 0          ; enable PORTB pull-ups to make B4-B7
                          ; high
TRISB = %11110000         ; make B7-B4 inputs, B3-B0 outputs
                          ; set up the initial LCD readings
LCDOUT $FE, 1             ; clear the LCD
LOOP:                     ;
  PORTB = %00001110       ; set line B0 low so we can read row
                          ; 1 only
  FOR ALPHA = 1 TO 4      ; need to look at 4 rows
    IF (PORTB & $F0)<>$F0 THEN ;
                          ; as soon as one of the bits in B4
                          ; to B7 changes we
                          ; immediately
                          ; have to store the value of PORTB in
                          ; a safe place.
    BUFFER = PORTB              ;
    GOSUB SHOWKEYPRESS          ;
    ELSE                        ;
    ENDIF                       ;
    PORTB = PORTB <<1           ; move bits left one place for
                                ; next line low
    PORTB = PORTB + 1           ; put 1 back in LSBit, the
                                ; right bit
  NEXT ALPHA                    ;
GOTO LOOP                       ;
                                ;
SHOWKEYPRESS:                   ;
  BUFFER = BUFFER ^ %11111111          ; reverses all the bits
                                       ; in the buffer
                                       ; print the first line
  LCDOUT $FE, $80, "ROW=",BIN4 (BUFFER & $0F)," COL=",
BIN4 BUFFER >>4
  COLUMN = (NCD BUFFER) -4             ; calculate column
  ROW = NCD (BUFFER &$0F)              ; calculate row
  SWITCH = ((ROW-1) * 4) +COLUMN       ; calculate switch number
                                       ; print the second line
  LCDOUT $FE, $C0, "ROW=", DEC ROW, " COL=", DEC COLUMN,
" SW=", DEC SWITCH, " "                ;
RETURN                                 ;
END                                    ;
```

Read one potentiometer and display its 8 bit value on LCD in binary, hex, and decimal notation, and impress the binary value on the LED bar graph.
A detailed discussion of A to D conversions is covered in the data sheets as mentioned earlier in the chapter.

As mentioned before, the potentiometers are read by dividing the voltage across a potentiometer into 256 parts and seeing which of the 256 divisions match the position of the wiper. This gives a reading between 0 and 255 (in 8-bit resolution). It does not tell us anything about the resistance of the potentiometer, only the relative position of the wiper.

We will read/use the pot closest to the edge of the board. This pot is connected to line PORTA.0 which is pin 2 of the MCU.

A to D conversions are controlled by the ADCON0 and ADCON1 registers and the 16F877A has to be in analog mode for the relevant bit for the A to D conversion to be enabled.

Setting the bits in ADCON0. (See data sheet for more information.)

Bits 7 and 6 Control the clock/oscillator to be used. Set these both to 1.
Bits 5 to 3 Select which channels are to be used in the conversions, Set to 000 for PORTA.0.
Bit 2 Cleared when the conversion is completed. Set it to 1 to start the conversion.
Bit 1 Ignored in A/D conversions. Set to 0.
Bit 0 Controls A/D conversions. Set to 1 to enable A/D conversions.

When the conversion is completed, the result will be placed in ADRESH and ADRESL. The format of how this is done depends on how the result is set up with register ADCON1.

ADCON1 needs bit 7 to be set to 0 to make the 8-bit result appear in ADRESH, and bit 2 needs to be set to 1 to select potentiometer 0 and set the proper reference voltages. See pages from the data sheet.

So we set ADCON0 to %11000001 to set up for reading PORTA.0, and we set ADCON1 to %00000010.

The program segment to read a value is as follows:

```
LOOP:                                    ; begin loop
  ADCON0.2 = 1                           ; start conversion
  NOT_DONE:                              ; marker if not done
    PAUSE 5              ;
  IF ADCON0.2 = 1 THEN NOT_DONE          ; wait for low on bit-2
                                         ; of ADCON0,
                                         ; conversion finished
  A2D_VALUE = ADRESH                     ; move high byte of
                                         ; result to A2D_Value
  LCDOUT $FE, 1                          ; clear screen
  LCDOUT "VALUE: ", DEC A2D_VALUE," "        ; display the
                                             ; decimal value
  PAUSE 100                              ; wait 0.1 second
GOTO LOOP                                ; do it forever
```

The complete program is listed in Program 5.16.

Program 5.16 Potentiometer readings. Display the value of potentiometer in all formats.

```
CLEAR                                  ; clear memory
DEFINE OSC 4                           ; osc speed
DEFINE LCD_DREG PORTD                  ; define LCD registers and bits
DEFINE LCD_DBIT 4                      ;
DEFINE LCD_RSREG PORTE                 ;
DEFINE LCD_RSBIT 0                     ;
DEFINE LCD_EREG PORTE                  ;
DEFINE LCD_EBIT 1                      ;
A2D_VALUE VAR BYTE                     ; create A2D_Value to store
                                       ; result
                                       ;
TRISA = %11111111                      ; set PORTA to all input
TRISD = %00000000                      ; set PORTD to all output
ADCON0 = %11000001                     ; configure and turn on A/D
                                       ; Module
ADCON1 = %00000010                     ; set PORTA analog and LEFT
                                       ; justify result
PAUSE 500                              ; wait 0.5 second for LCD
                                       ; startup
                                       ;
LOOP:                                  ;
  ADCON0.2 = 1                         ; start conversion
  NOT_DONE:                            ;
  IF ADCON0.2 = 1 THEN NOT_DONE ; wait for low on bit-2 of
                                       ; ADCON0,
                          ; conversion finishes
  A2D_VALUE = ADRESH      ; move high byte of result to
                          ; A2D_Value
  LCDOUT $FE, 1           ; clear screen
  LCDOUT "DEC VALUE = ", DEC A2D_VALUE," " ; Display 3 values
  LCDOUT $FE, $C0, "HEX=", HEX2 A2D_VALUE," ","BIN=", BIN8
A2D_VALUE," "
  PORTD =A2D_VALUE        ; displays value in bar graph
  PAUSE 100               ; wait 0.1 second
GOTO LOOP                 ; do it forever
END                       ; end program
```

Program 5.16 uses the named registers themselves to set up the conversions. In the next program, we will use the power of the compiler and its related commands to read the three pots much more conveniently with the ADCIN command.

Read all three potentiometers and display their values on the LCD using the ADCIN command.

Five of six of the available pins on PORTA can be used as analog inputs. In our case pins 0, 1, and 3 are connected to the three potentiometers.

If we want to read all three pots, we have to activate their three lines and create variables to store the three results that are obtained. The modifications to Program 5.16 are implemented in Program 5.17.

Program 5.17 Display potentiometer settings. Read and display all three potentiometers values in decimal format.

```
CLEAR                        ; clear memory
DEFINE OSC 4                 ; osc speed
DEFINE LCD_DREG PORTD        ; define LCD connections
DEFINE LCD_DBIT 4            ;
DEFINE LCD_RSREG PORTE       ;
DEFINE LCD_RSBIT 0           ;
DEFINE LCD_EREG PORTE        ;
DEFINE LCD_EBIT 1            ;
LOW PORTE.2                  ; LCD R/W line low (W)
PAUSE 500                    ; wait .5 second for LCD
                             ; the next 3 defines are needed for
                             ; the ADCIN command
DEFINE ADC_BITS 8            ; set number of bits in result
DEFINE ADC_CLOCK 3           ; set internal clock source (3=rc)
DEFINE ADC_SAMPLEUS 50       ; set sampling time in µS
                             ;
TRISA = %11111111            ; set PORTA to all input
TRISD = %00000000            ; set all PORTD lines to outputs
ADCON1 = %00000110           ; PORTA and PORTE to digital
A2D_Value0 VAR BYTE          ; Create A2D_Value to store result 1
A2D_Value1 VAR BYTE          ; Create A2D_Value to store result 2
A2D_Value2 VAR BYTE          ; Create A2D_Value to store result 3
LCDOUT $FE, 1                ; clear the display
                             ;
MAINLOOP:                    ; main program loop
                             ; read in the potentiometer values
  ADCIN 0, A2D_VALUE0        ; read channel 0 to A2D_Value0
  ADCIN 1, A2D_VALUE1        ; read channel 1 to A2D_Value1
  ADCIN 3, A2D_VALUE2        ; read channel 2 to A2D_Value2
  LCDOUT $FE, $80, DEC A2D_VALUE0," ",DEC A2D_VALUE1," " ,_
; DEC A2D_VALUE2," ";
  PAUSE 10                   ;
GOTO MAINLOOP                ; do it all forever
END                          ; end program
```

Adding the kind of flexibility that defines computer interfaces and demonstrates the ability to make sophisticated real-time adjustments:

- Use the three potentiometers to control one R/C servo.
- Control the relative location of the center position with POT1.

- Control limit position of the end positions with POT2.
- Control the rate of movement with POT3.

Program 5.18 Servo/Potentiometers: three potentiometers controlling one servo. Connect the servo to Jumper J7 for this program.

```
CLEAR                             ; clear memory
DEFINE OSC 4                      ; osc speed
DEFINE LCD_DREG PORTD             ; define LCD connections
DEFINE LCD_DBIT 4                 ;
DEFINE LCD_RSREG PORTE            ;
DEFINE LCD_RSBIT 0                ;
DEFINE LCD_EREG PORTE             ;
DEFINE LCD_EBIT 1                 ;
LOW PORTE.2                       ; LCD R/W line low (W)
DEFINE ADC_BITS 8                 ; set number of bits in result
DEFINE ADC_CLOCK 3                ; set clock source (3=rc)
DEFINE ADC_SAMPLEUS 50            ; set sampling time in µS
TRISA        = %11111111          ; set PORTA to all input
TRISD        = %00000000          ; set all PORTD lines to
                                  ; outputs
ADCON1 = %00000111                ; PORTA and PORTE to digital
A2D_VALUE    VAR BYTE             ; create A2D_Value to store
                                  ; result
A2D_VALUE1 VAR BYTE               ; create A2D_Value1 to store
                                  ; result
A2D_VALUE2 VAR BYTE               ; create A2D_Value2 to store
                                  ; result
ADWALWAS VAR BYTE                 ;
POS  VAR WORD                     ; servo positions
CENTERPOS VAR WORD                ;
MAXPOS      VAR WORD              ;
MINPOS      VAR WORD              ;
POSSTEP     VAR BYTE              ;
PAUSE 500                         ; wait 0.5 second
SERVO1      VAR PORTC.1           ; alias servo pin
ADCIN 0, A2D_VALUE                ; read channel 0 to A2D_Value
OPTION_REG = $01111111            ; enable PORTB pullups
LOW SERVO1                        ; servo output low
GOSUB CENTER                      ; center servo
LCDOUT $FE, 1                     ; clears screen only
PORTB = 0                         ; PORTB lines low to read
                                  ; buttons
TRISB = %11111110                 ; enable first button row
                                  ; main program loop
```

(continued)

Program 5.18 Servo/Potentiometers: three potentiometers controlling one servo. Connect the servo to Jumper J7 for this program. (*continued*)

```
MAINLOOP:                        ; check any button pressed to
                                 ; move servo
  IF PORTB.4 = 0 THEN GOSUB LEFT    ;
  IF PORTB.5 = 0 THEN GOSUB CENTER  ;
  IF PORTB.6 = 0 THEN GOSUB RIGHT   ;
  ADCIN 0, A2D_VALUE             ; read channel 0 to A2D_Value
  ADCIN 1, A2D_VALUE1            ; read channel 1 to A2D_Value 1
  ADCIN 3, A2D_VALUE2            ; read channel 2 to A2D_Value 2
  MAXPOS = 1500 + A2D_VALUE1*3      ;
  MINPOS = 1500 - A2D_VALUE1*3      ;
  CENTERPOS = 1500+3*(A2D_VALUE-127);
  POSSTEP = A2D_VALUE2/10 +1     ;
  SERVO1 = 1                     ; start servo pulse
  PAUSEUS POS                    ;
  SERVO1 = 0                     ; end servo pulse
  LCDOUT $FE, $80, "POS=", DEC POS , "  "` ;
  LCDOUT $FE, $C0, DEC A2D_VALUE," ",DEC A2D_VALUE1," ",
DEC POSSTEP," "
  PAUSE 10                       ; servo update rate about 60 Hz
GOTO MAINLOOP                    ; do it all forever
                                 ;
LEFT:                            ; move servo left
IF POS < MAXPOS THEN POS = POS + POSSTEP
RETURN                           ;
                                 ;
RIGHT:                           ; move servo right
IF POS > MINPOS THEN POS = POS - POSSTEP
RETURN                           ;
                                 ;
CENTER:                          ; center servo
POS = 1500+3*(A2D_VALUE-127)     ;
RETURN                           ;
END                              ; end program
```

Exercises

Answers to these problems are not provided.

Since this is really all about input and output, a comprehensive set of exercises that focus specifically on input and output have been provided. We need to be completely comfortable with these I/O functions before we start on running motors, so you are encouraged to expand on these exercises on your own. These and similar techniques will be used to control and respond to all the ancillary devices that we will use with our motors.

LED EXERCISES: CONTROLLING THE LIGHT EMITTING DIODES (LEDS)

We will learn more about controlling the output from the LAB-X1 by writing a series of increasingly complicated programs that will control the ten-segment LED IC provided on the LAB-X1. In these exercises we are controlling the LEDs, but the control strategies developed will apply to any kind of "on/off" devices that we will connect to the LAB-X1 or to any other device that we may design.

1. One at a time, light the eight LEDs on the right till they are all lit, and then turn them off one at a time. Time delay between actions is to be as close to one tenth of a second as you can get it.
2. Modify the preceding program so that the delay time is controlled by the topmost potentiometer on the LAB-X1. The time is to vary from 10 milliseconds to 200 milliseconds—no less, no more.
3. Write a program that will vary the glow on the rightmost LED from fully off to fully on once a second. Program the second LED to go dark and bright exactly 180 degrees out of phase with the first LED so that as one LED is getting brighter, the other LED gets dimmer, and vice versa.
4. Write a program that flashes the four leftmost LEDs on and off every 0.25 seconds and cycles the four LEDs on the right through a bright/dim cycle every two seconds.
5. Write a program that flashes the first LED 10 times a second, flashes the second one 9 times a second, and flashes the third LED whenever both LEDs are on at the same time. Display how many times the third LED has blinked on the LCD display. (Timing can be approximate but has to have a common divider so the third LED will give the beat frequency.)

LIQUID CRYSTAL DISPLAY EXERCISES: CONTROLLING THE LCD

The addresses of the memory locations used by the LCD have already been fixed, as has the instruction set that we use to write to the LCD. The description of the Hitachi HD44780U (LCD-II) controller instruction set, as well as its electronic characteristics, are given in the data sheet provided for the display. Here we will list only the codes that apply to our immediate use of the device.

There are two types of commands that can be sent to the display: the control codes and the set of actual characters to be displayed. Both uppercase and lowercase characters are supported, as are a number of special and graphic characters. The control codes allow you to control the display and set the position of the cursor and so on.

Each control code has to be preceded by decimal 254 or HEX$FE. (The controller also supports the display of a set of Japanese characters, which are not of interest to us.)

Command codes for the following actions are provided along with others. Go to the data sheet for the controller to learn what all these command codes are.

- Clear the LCD
- Return home
- Go to beginning of line 1
- Go to beginning of line 2
- Go to a specific position on line 1
- Go to a specific position on line 2
- Show the cursor
- Hide the cursor
- Use an underline cursor
- Turn on cursor blink
- Move cursor right one position
- Move cursor left one position

There are still other commands that you will discover in the data sheet. There are more memory locations within the LCD, and there are invisible locations beyond the end of the visible 20 characters.

It is also possible to design your own font for use with this particular display; all the information you need to do so is in the Hitachi HD44780U book/data sheet.

1. Write a program to put the 26 letters of the alphabet and the 10 numerals in the 40 spaces that are available on the display. Put four spaces between the numbers and the alphabet to fill in the four remaining spaces. Once all the characters have been entered, scroll the 40 characters back and forth endlessly though the two lines of the display.
2. Write a program to bubble the 26 capital letters of the alphabet through the numbers 0 to 9 on line two of the LCD. To do this, first put the numbers on line two. Then "A" takes the place of the "0" and all the numbers move over. Then the "A" takes the place of the "1" and the "0" moves to position 1. Then the "A" moves into place of the "2," and so on till it gets past the 9. Then the "B" starts its way across the numbers and so on. Loop forever.
3. Write a program to write the numbers 0 to 9 upside down on line 1. Wait one second and then flip the numbers right side up. Loop.
4. Write a program to write "HELLO WORLD" to the display and then change it to lowercase one letter at a time with 50 milliseconds between letters. Wait one second and go back to uppercase one character at a time with negative letters (all dots on the display are reversed to show as dark background with white letters and on to lowercase). Loop.

ADVANCED EXERCISES

These exercises are designed to challenge your programming ability. Again, you will need access to the data sheet for the LC Display.

1. Editor: Write a program that displays 12 random numbers on line 1 of the LCD and displays a cursor that can be moved back and forth across the 20 spaces with potentiometer 0. The entire range of the potentiometer must be used to move across the 20 spaces. Allow the keypad to insert numbers 0 to 9 into the position that the cursor is on. Assign a delete switch and an insert space switch on the keyboard. A comprehensive number (plus decimal and space) editor is required.
2. Mirror: Write a program that puts a random set of letters and numbers on line 1 and then puts their mirror images on line 2. The mirror is between line 1 and line 2. To do this, you have to learn how to create the upside down numbers from the Hitachi data sheet for the display, and you have to learn how to read what is in the display from the display ROM.
3. Forty characters: The display ROM is capable of storing 40 characters on each line. Design a program to allow you to scroll back and forth to see all 40 characters on both lines one line at a time. Use two potentiometers for scrolling, one for each line.
4. Four lines: Write a program to display four lines of random data on the LCD and to scroll up and down and side to side to see all four lines in their entirety. You have to store what is lost from the screen before it is lost so that you can re-create it when you need it.
5. Bar graphs: Create a three-bar graph display, with each bar 3 pixels high, that extends across both lines of the LCD. The lengths of the bar graphs are determined by the settings of the three potentiometers and changes as the potentiometers are manipulated.

6

TIMERS AND COUNTERS

If you have no knowledge about timers, you should read this chapter a few times. However, there is some repetition in the other chapters to allow each part of the book to stand as an independent resource.

Most users will find that using the timers and the counters is the hardest part of learning how to use PIC microcontrollers. With this in mind, we will proceed in a step-by-step manner and build up the programs in pieces that are easier to digest. Once you get comfortable with their setup procedures, you will find that timers and counters are not so intimidating.

We will cover timers and counters separately. Counters are essentially timers that get their clock input from an outside source. There are two counters in the 16F877A, and they are associated with Timer0 and Timer1. Timer2 cannot be used as a counter because there is no input line (internal or external) for this particular counter.

Note *The clock frequency utilized by the timers is a fourth of the oscillator frequency. This is the frequency of the instruction clock. This means that the counters are affected by every fourth count of the main oscillator. The frequency is referred to as Fosc/4 in the data sheet. When responding to an external clock signal the response is to the actual frequency of the external input. (For now, we'll run the LAB-X1 and thus all the programs at 4 MHz, so all the timers will be getting internal inputs at 1 MHz.)*

Caution *The PICBASIC PRO compiler generates code that does not respond to interrupts while a compiled instruction is being executed. Therefore long pauses (long enough to lose an interrupt signal, which depends on how the timer is set up) can lead to lost interrupts if more than one interrupt occurs during the pause. Since interrupts are used for the express purpose of handling critical response/timing needs, this is most undesirable. Therefore, PAUSE commands should be used with care. The program samples provided in this chapter give examples of how to use short pauses in loops to get a long pause.*

Timers

Timer0 will be covered first, and in more detail, as a prototypical timer, and discussion and examples for the use of Timer1 and Timer2 will be provided.

The use of timers internal to microprocessors is a bit more complicated than what we have been doing so far because there is a considerable amount of setup required before the timer can be used, and the options for setting the timers up are extensive. We will cover the timers one at a time in an introductory manner. However, you should be aware that there is an entire book available from Microchip Technologies that covers nothing but timers, so the coverage here will be rudimentary.

Note *The Microchip Technologies timer manual is called* The PICmicro Mid-Range MCU Family Reference Book *(DS33023), available from Microchip Technology Inc.*

To understand timers, you must understand how to turn them on and off and how to read and set the various bits and bytes that relate to them. Essentially, in the typical timer application you turn on a timer by turning on its enable bit. The timer then counts a certain number of clock cycles, sets an interrupt bit that causes an interrupt, and continues running toward the next interrupt. Your program responds to the interrupt by executing a specific interrupt handling routine and then clearing the interrupt bit. The program then returns to wherever it was when the interrupt occurred. The pre-/post-scalers have to do with modifying the time it takes for an interrupt to take place. The hard part is getting familiar with which bit does what and where it is located, which is why reading and understanding the data sheet chapters (5, 6, or 7) depending on the timer you are using is imperative. There is no escaping this horror! However, this chapter will ease your pain.

Timers allow the microcontroller to create and react to chronological events. These include:

- Timing events for communications
- Creation of clocks for various purposes
- Generating timed interrupts
- Controlling PWM generation
- Waking the PIC up from the sleep mode at intervals to do work and go back to sleep
- Special use of the Watchdog Timer

There are three internal timers in the PIC 16F877A (there is also a Watchdog Timer, which is discussed after the timers in this chapter):

- Timer0 is an 8-bit free running timer/counter with an optional pre-scaler. It is the simplest of the timers to use.
- Timer1 is a 16-bit timer that can be used as a timer or as a counter. It is the only 16-bit timer and can also be used as a counter. It is the most complicated of the timers.

- Timer2 is an 8-bit timer with a pre-scaler and a post-scaler and cannot be used as a counter. (There is no way to input a signal to this timer.)

Each timer has a timer control register that sets the options and properties that the timer will exhibit. All the timers are similar, but each of them has special features that give it special properties. It is imperative that you refer to the data sheet for the PIC 16F877A (Chapters 5, 6, and 7) as you experiment with the timer functions. Once you start to understand what the PIC designers are up to with the timer functions, it will start to come together in your mind.

Timers can have pre-scalers and post-scalers associated with them that can be used to multiply the timer setting by a limited number of integer counts. The scaling ability is not adequate to allow all exact time intervals to be created, but it is adequate for all practical purposes. To the inability to create perfectly timed interrupts we have to add the uncertainty in the frequency of the oscillator crystal, which is usually not exactly what it is stated to be (and which is affected by the ambient temperature as the circuitry warms up). Though fairly accurate timings can be achieved with the hardware as received, additional software adjustments may have to be added if extremely accurate results are desired. The software can be designed to make a correction to the timing every so often to make it more accurate. We will also need to have an external source that is at least as accurate as we want our timer to be so we can verify the accuracy of the device that we create.

TIMER0

First we'll write a simple program to see how this timer works. We will use the LED bar graph to show what is going on inside the microcontroller.

As always, the bar graph is connected to the eight lines of PORTD of the LAB-X1.

First, we will write a program that will light the two LEDs connected to D0 and D1 alternately. Having them light alternately lets you know that the program is running or, more accurately, the segment of the program that contains this part of the code is running. These two LEDs will be used to represent the foreground task in the program. There is no timer process in use in Program 6.1 at this stage.

Program 6.1 Foreground program blinks two LEDs alternately. No timer is being used at this time.

```
CLEAR                     ; clears all memory locations
DEFINE OSC 4              ; using a 4 MHz oscillator here
TRISD = %11110000         ; make D0 to D3 outputs, rest inputs
PORTD.0 = 0               ; turn off bit D0
PORTD.1 = 1               ; turn on bit D1
ALPHA VAR WORD            ; set up a variable for counting
                          ;
MAINLOOP:                 ; main loop
```

(continued)

Program 6.1 Foreground program blinks two LEDs alternately. No timer is being used at this time. (*continued*)

```
  IF PORTD.1 = 0 THEN   ; the next lines of code turn the
                        ; LEDs on
    PORTD.1 = 1         ; if they are off
    PORTD.0 = 0         ;
  ELSE                  ; and off if they
    PORTD.1 = 0         ; are on
    PORTD.0 = 1         ;
  ENDIF                 ;
  FOR ALPHA = 1 TO 300  ; this loop replaces a long pause
                        ; command
    PAUSEUS 100         ; with short pauses that are
                        ; essentially
  NEXT ALPHA            ; independent of the clock frequency.
GOTO MAINLOOP           ; do it all forever
END                     ; all programs need to end with END
```

The use of the PAUSEUS loop in Program 6.1 provides a latency of 100 μs (worst case) in the response to an interrupt and eliminates most of the effect of changing the OSC frequency if that should become necessary. It is better than using an empty counter, which would be completely dependent of the frequency of the system oscillator. There is an assumption here that the 100 microsecond latency is completely tolerable to the task at hand and it is for this program. This may not be true for your real-world program and may have to be adjusted.

We are turning one LED off and another LED on to provide a more positive feedback. As long as we are executing the main loop, the LCDs will light alternately and provide a dynamic feedback of the operation of the program in the foreground loop. It is important that you learn to develop failsafe techniques for your programs, and this is a rudimentary one for making sure that a program is running.

We have selected a relatively fast on/off cycle so that we will better be able to see minor delays and glitches that may appear in the operation of the program as we proceed.

Run this program to get familiar with the operation of the two LEDs. Adjust the counter (the 300 value) to suit your taste.

Next, in Program 6.2, we will add the code that will interrupt this program periodically and make a third LED go on and then off on an approximately one-second cycle. This will serve as the interrupt driven task that we are interested in learning how to create. In most programs this would be the critical, time-dependent task.

Here is what you have to do to make the interrupt driven LED operational:

- Enable Timer0 and its interrupts with appropriate register/bit settings.
- Add the ON INTERRUPT command to tell the program where to go to handle the interrupt when an interrupt occurs.
- Set up the interrupt routine to do what needs to be done (the interrupt routine counts to 61 and turns the LED on if it is off and off if it is on).

- Clear the interrupt flag that was set by Timer0.
- Send the program back to where it was interrupted with the RESUME command.

WHY ARE WE USING 61?

- Set the pre-scaler to 64 (bits 0 to 2 are set at 101 in the OPTION_REG).
- Set the counter to interrupt every 256 counts ($256 \times 64 = 16,384$).
- Set the clock to 4,000,000 Hz
- Set Fosc/4 to 1,000,000 (1,000,000 / 16,384 = 61.0532—it's not exact but close enough for our purposes for now).

The lines of code now look like Program 6.2.

Program 6.2 **Using Timer0.** Blinks two LEDs (D1 and D0) alternately and blinks a third LED (D2) for one second on and one second off as controlled by the interrupt signal.

```
CLEAR                         ; clear memory
DEFINE OSC 4                  ; using a 4 MHz oscillator
                              ; set the option register
OPTION_REG = %10000101        ; page 48 of data sheet
                              ; bit 7=1 disables pull-ups on PORTB
                              ; bit 5=0 selects timer mode
                              ; bit 2=1 }
                              ; bit 1=0 } sets Timer0 pre-scaler to 64
                              ; bit 0=1 }
                              ; sets the interrupt control register
INTCON = %10100000            ; bit 7=1 enables all unmasked
                              ; interrupts
                              ; bit 5=1 enables Timer0 overflow
                              ; interrupt
                              ; bit 2 flag will be set on interrupt
                              ; and has to be cleared
                              ; in the interrupt routine. It is set
                              ; clear to start with.
ALPHA VAR WORD                ; this variable counts in the Pause µS
                              ; loop
BETA VAR BYTE                 ; this variable counts the 61
                              ; interrupt ticks
TRISD = %11110100             ; sets the 3 output pins in the D port
PORTD = %00000000             ; sets all pins low in the D port
BETA =0                       ; sets the counter to zero
ON INTERRUPT GOTO INTERUPTROUTINE   ; this line needs to be
                                    ; early in the program,
                              ; in any case, before the routine is
                              ; called.
                              ;
MAINLOOP:                     ; main loop blinks D0 and D1
                              ; alternately
```

(continued)

Program 6.2 Using Timer0. Blinks two LEDs (D1 and D0) alternately and blinks a third LED (D2) for one second on and one second off as controlled by the interrupt signal. (*continued*)

```
  IF PORTD.1 = 0 THEN    ; ]
    PORTD.1 = 1          ; ]
    PORTD.0 = 0          ; ] this part of the program blinks
                         ; two LEDs in
  ELSE                   ; ] the foreground as described
                         ; before
    PORTD.1 = 0          ; ]
    PORTD.0 = 1          ; ]
  ENDIF                  ; ]
                         ;
  FOR ALPHA = 1 TO 300   ; the long pause is eliminated with
                         ; this loop
    PAUSEUS 100          ; PAUSE command with short latency
  NEXT ALPHA             ;
GOTO MAINLOOP            ; end of loop
                         ;
DISABLE                  ; DISABLE and ENABLE must bracket
                         ; interrupt routine
INTERUPTROUTINE:         ; this information is used by the
                         ; compiler only
  BETA = BETA + 1        ;
  IF BETA < 61 THEN ENDINTERRUPT    ; one second has not yet
                                    ; passed
  BETA = 0               ; reset the counter after it overflows
  IF PORTD.3 = 1 THEN    ; interrupt loop turns D3 on and off
                         ; every
    PORTD.3 = 0          ; 61 times through the interrupt
                         ; routine
  ELSE                   ; That is about one second per full
                         ; cycle
    PORTD.3 = 1          ;
  ENDIF                  ;
ENDINTERRUPT:            ; used if one sec has not elapsed
  INTCON.2 = 0           ; clears the interrupt flag for this
                         ; timer
RESUME                   ; resume the main program
ENABLE                   ; DISABLE and ENABLE must bracket the
                         ; int. routine
END                      ; end program
```

Make your predictions and then try changing the three low bits in OPTION_REG to see how they affect the operation of the interrupt.

In Program 6.2, Timer0 is running free and providing an interrupt every time its 8 bit counter overflows from FF to 00. The pre-scaler is set to 64 so we get the interrupt after 64 of these interrupt cycles. When this happens we jump to the "InterruptRoutine" routine, where we make sure that 61 interrupts have taken place. If they have, we change the state of an LED and return to the place where the interrupt took place. (It happens that it takes approximately 61 interrupts to equal one second in this routine with a processor running at 4 MHz. This could be refined by trial and error after initial calculation if necessary.)

Note that the interrupts are disabled while we are in the "InterruptRoutine" routine, but the free running counter is still running toward its next overflow, meaning that whatever we do has to get done in less than 1/61 seconds if we are not going to miss the next interrupt (unless we make some other arrangements to count all the interrupts with an internal subroutine or some other scheme). It can become quite complicated and we will not worry about it for now.

Before going any further, let's take a closer look at the OPTION_REG and the INTCON (INTerrupt CONtrol) register. These registers were used in Program 6.1, but the details of their operations were not explained. These are 8-bit registers with the eight bits of each register assigned as follows:

OPTION_REG the option register

Bit 7 RBPU. Not of interest to us at this time. (This bit enables the port B weak pull-ups.)

Bit 6 INTEDG. Not of interest to us at this time. Interrupt edge select bit deter mines which edge the interrupt will be selected on, rising (1) or falling (0). Either one works for us.

Bit 5 TOCS, Timer0 Clock Select bit. Selects which clock will be used.
1 = Transition on TOCKI pin.
0 = Internal instruction cycle clock (CLKOUT). We will use this, the oscillater. See bit 4 description.

Bit 4 TOSE, Source Edge Select Bit. Determines when the counter will increment.
1 = Increment on high to low transition of TOCKI pin.
0 = Increment on low to high transition of TOCKI pin.

Bit 3 PSA, Pre-scaler assignment pin. Decides what the pre-scaler applies to.
1= Select Watch Dog Timer (WDT).
0 = Select Timer0. We will be using this.

Bits 2, 1 and 0 define the pre-scaler value for the timer. As mentioned previously, the pre-scaler can be associated with Timer0 or with the Watchdog Timer (WDT) but not both. Note that the scaling for the WDT is half the value for Timer0 for the same three bits.

Bit value	TMR0 rate	WDT rate
000	1:2	1:1
001	1:4	1:2
010	1:8	1:4

(*continued*)

Bit value	TMR0 rate	WDT rate
011	1:16	1:8
100	1:32	1:16
101	1:64	1:32 (we will use this)
110	1:128	1:64
111	1:256	1:128

Caution *There is a very specific sequence that must be followed (which does not apply here) when changing the pre-scaler assignment from Timer0 to the WDT (Watch Dog Timer) to make sure that an unintended reset does not take place. This is described in detail in* The PICmicro Mid-Range MCU Family Reference Book *(DS33023).*

As per the preceding list, for our specific example OPTION_REG is set to %10000101. (Refer to the data sheet for more specific information.)

INTCON the interrupt control register values are as follows.

Bit 7 = 1 Enables global interrupts. If you are going to use interrupts this bit must be set.
Bit 6 = 1 Enables all peripheral interrupts.
Bit 5 = 1 Enables an interrupt to be set on Bit 2 below when Timer0 overflows.
Bit 4 = 1 Enables an interrupt if RB0 changes.
Bit 3 = 1 Enables an interrupt if any of the PORTB pins is programmed as input and change state.
Bit 2 Is the Interrupt flag for Timer0.
Bit 1 Is the Interrupt flag for all internal interrupts.
Bit 0 Is the Interrupt flag for pins B7 to B4 if they change state.

Note that Bit 2 is set clear when we start and will be set to 1 when the interrupt takes place. It has to be recleared within every interrupt service routine, usually at the end of the interrupt routine.

A Timer0 Clock

The following program (Program 6.3) written by microEngineering Labs and provided by them as a part of the information on their web site demonstrates the use of interrupts to create a reasonably accurate clock that uses the LCD display to show the time in hours, minutes, and seconds. (I did not modify this program in any way so it does not include the CLEAR or OSC statements and so on that we have been using in our programs.)

```
; LCD clock program using On Interrupt
; Uses TMR0 and pre-scaler. Watchdog Timer should be
; set to off at program time and Nap and Sleep should not be
; used.
; Buttons may be used to set hours and minutes.
```

Program 6.3 Timer0 usage per microEngineering Labs program

```
hours, seconds and minutes digital clock
DEFINE LCD_DREG PORTD  ; define LCD connections
DEFINE LCD_DBIT 4      ;
DEFINE LCD_RSREG PORTE ;
DEFINE LCD_RSBIT 0     ;
DEFINE LCD_EREG PORTE  ;
DEFINE LCD_EBIT 1      ;
                       ;
HOUR VAR BYTE          ; define hour variable
DHOUR VAR BYTE         ; define display hour variable
MINUTE VAR BYTE        ; define minute variable
SECOND VAR BYTE        ; define second variable
TICKS VAR BYTE         ; define pieces of seconds variable
UPDATE VAR BYTE        ; define variable to indicate update
                       ; of LCD
I VAR BYTE             ; debounce loop variable
ADCON1 = %00000111     ; parts of PORTA and E made digital
LOW PORTE.2            ; LCD R/W low = write
PAUSE 100              ; wait for LCD to startup
HOUR = 0               ; set initial time to 00:00:00
MINUTE = 0             ;
SECOND = 0             ;
TICKS = 0              ;
UPDATE = 1             ; force first display
                       ; set TMR0 to interrupt every 16.384
                       ; ms
OPTION_REG = %01010101 ; set TMR0 configuration and enable
                       ; PORTB pullups
INTCON = %10100000     ; enable TMR0 interrupts
ON INTERRUPT GOTO TICKINT    ;
                       ; main program loop -
MAINLOOP:              ; in this case, it only updates the
                       ; LCD
TRISB = %11110000      ; enable all buttons
PORTB =%00000000       ; PORTB lines low to read buttons
                       ; check any button pressed to set time
IF PORTB.7 = 0 THEN DECMIN   ;
IF PORTB.6 = 0 THEN INCMIN   ; last 2 buttons set minute
IF PORTB.5 = 0 THEN DECHR    ;
IF PORTB.4 = 0 THEN INCHR    ;
                             ; first 2 buttons set hour
CHKUP: IF UPDATE = 1 THEN    ; check for time to update
                             ; screen
LCDOUT $FE, 1                ; clear screen
```

(continued)

Program 6.3 **Timer0 usage per microEngineering Labs program** (*continued*)

```
                                ; display time as hh:mm:ss
 DHOUR = HOUR                   ; change hour 0 to 12
 IF (HOUR // 12) = 0 THEN       ;
      DHOUR = DHOUR + 12        ;
 ENDIF                          ;
                                ;
 IF HOUR < 12 THEN              ; check for AM or PM
  LCDOUT DEC2 DHOUR, ":", DEC2 MINUTE, ":", DEC2 second, " AM"
  ELSE                          ;
  LCDOUT DEC2 (DHOUR - 12), ":", DEC2 MINUTE, ":", DEC2
SECOND, " PM"
  ENDIF                         ;
  UPDATE = 0                    ; screen updated
 ENDIF                          ;
 GOTO MAINLOOP                  ; do it all forever
                                ; Increment minutes
 INCMIN: MINUTE = MINUTE + 1    ;
 IF MINUTE >= 60 THEN           ;
  MINUTE = 0                    ;
 ENDIF                          ;
 GOTO DEBOUNCE                  ;
                                ; increment hours
 INCHR: HOUR = HOUR + 1         ;
 IF HOUR >= 24 THEN             ;
  HOUR = 0                      ;
 ENDIF                          ;
 GOTO DEBOUNCE                  ;
                                ; decrement minutes
 DECMIN: MINUTE = MINUTE - 1    ;
 IF MINUTE >= 60 THEN           ;
  MINUTE = 59                   ;
 ENDIF                          ;
 GOTO DEBOUNCE                  ;
                                ; decrement hours
 DECHR: HOUR = HOUR - 1         ;
 IF HOUR >= 24 THEN             ;
  HOUR = 23                     ;
 ENDIF                          ;
                                ; debounce and delay for
                                ; 250 ms
 DEBOUNCE: FOR I = 1 TO 25      ;
  PAUSE 10                      ; 10 ms at a time so no
                                ; interrupts are lost
 NEXT I                         ;
 UPDATE = 1                     ; set to update screen
```

(*continued*)

Program 6.3 **Timer0 usage per microEngineering Labs program** (*continued*)

```
GOTO CHKUP                      ;
                                ; Interrupt routine to handle
                                ; each timer tick
DISABLE                         ; disable interrupts during
                                ; interrupt handler
  TICKINT: TICKS = TICKS + 1    ; count pieces of seconds
    IF TICKS < 61 THEN TIEXIT   ; 61 ticks per second (16.384
                                ; ms per tick)
                                ; one second elapsed – update
                                ; time
    TICKS = 0                   ;
    SECOND = SECOND + 1         ;
      IF SECOND >= 60 THEN      ;
        SECOND = 0              ;
        MINUTE = MINUTE + 1     ;
      IF MINUTE >= 60 THEN      ;
        MINUTE = 0              ;
        HOUR = HOUR + 1         ;
          IF HOUR >= 24 THEN    ;
          HOUR = 0              ;
        ENDIF                   ;
      ENDIF                     ;
    ENDIF                       ;
  UPDATE = 1                    ; Set to update LCD
TIEXIT: INTCON.2 = 0            ; reset timer interrupt flag
RESUME                          ;
END                             ;
```

In the clock implemented in Program 6.3, the keyboard buttons are used as follows:

- SW1 and SW5 increment the hours
- SW2 and SW6 decrement the hours
- SW3 and SW7 increment the minutes
- SW4 and SW8 decrement the minutes

The seconds cannot be affected other than with the reset switch.

TIMER1: THE SECOND TIMER

The second timer, Timer1, is the 16 bit-timer/counter. This is the most powerful timer in the MCU. As such it is the hardest of the three timers to understand and use, and it is also the most flexible. It consists of two 8-bit registers; each register can be read and be written to. The timer can be used either as a timer or as a counter depending on how the Timer1 Clock Select bit (TMR1CS) is set. This bit is Bit 1 in the Timer1 Control Register (T1CON).

In Timer1, we can set the value the timer starts its count with, and thus change the frequency of the interrupts. Here we are looking to see the effect of changing the value preloaded into Timer1 on the frequency of the interrupts as reflected in a very rudimentary pseudo stopwatch. The higher the value of the preload, the sooner the counter will get to $FFFF, and the faster the interrupts will come. We will display the value of the pre-scaler loaded into the timer on the LCD so that we can see the correlation between the values and the actual operation of the interrupts. As the interrupts get closer and closer together, the time left to do the main task gets shorter and shorter and you can see this on the speed at which the stopwatch runs.

In Program 6.4 the switches perform the following actions:

- SW1 turns the stopwatch on
- SW2 stops the stopwatch
- SW3 resets the stopwatch

POT1 is the first potentiometer. It is read and then written to TMR1H to change the interrupt rate. (We are ignoring the low byte because it does not affect the interrupt rate much in this experiment.)

The results of the experiment are displayed on the LCD display.

Let us creep up on the solution. We will develop the program segments and discuss them as we go along and then put the segments together for a program we can run.

Program 6.4 Timer1 usage. Rudimentary timer operation which depends on value of POT1

```
; Set Up the LCD
CLEAR                        ; clear the memory
DEFINE OSC 4                 ; set the oscillator frequency
DEFINE LCD_DREG PORTD        ; LCD is on PORTD
DEFINE LCD_DBIT 4            ; we will use 4 bit protocol
DEFINE LCD_RSREG PORTE       ; register select register
DEFINE LCD_RSBIT 0           ; register select bit
DEFINE LCD_EREG PORTE        ; enable register
DEFINE LCD_EBIT 1            ; enable bit
PORTE.2 = 0                  ; set for write mode
PAUSE 500                    ; wait 0.5 second
                             ;
                             ; Next let us define the variables we
                             ; will be using
ADVAL        VAR    BYTE     ; Create adval to store result
TICKS        VAR    WORD     ;
TENTHS       VAR    BYTE     ;
SECS VAR     WORD            ;
MINS VAR     BYTE            ;
                             ;
```

(continued)

Program 6.4 Timer1 usage. Rudimentary timer operation which depends on value of POT1 (*continued*)

```
                                ; Set the variable to specific values,
                                ; not necessary
                                ; in this program but a formality for
                                ; clarity
TICKS = 0                       ;
TENTHS = 0                      ;
SECS = 0                        ;
MINS = 0                        ;
                                ; Set the registers that will control
                                ; the work. This is
                                ; the nitty gritty of it so we will
                                ; call out each bit.
                                ; INTCON is the interrupt control
                                ; register.
INTCON = %11000000
; bit 7: GIE: Global Interrupt Enable bit, this has to be set
; for any interrupt to work.
;        1 Enables all un-masked interrupts
;        0 = Disables all interrupts
; bit 6: PEIE: Peripheral Interrupt Enable bit
;        1 = Enables all unmasked peripheral interrupts
;        0 = Disables all peripheral interrupts
; bit 5: T0IE: TMR0 Overflow Interrupt Enable bit
;        1 = Enables the TMR1 interrupt
;        0 = Disables the TMR1 interrupt
; bit 4: INTE: RB0/INT External Interrupt Enable bit
;        1 = Enables the RB0/INT external interrupt
;        0 = Disables the RB0/INT external interrupt
; bit 3: RBIE: RB Port Change Interrupt Enable bit
;        1 = Enables the RB port change interrupt
;        0 = Disables the RB port change interrupt
; bit 2: T0IF: TMR0 Overflow Interrupt Flag bit
;        1 = TMR0 register has overflowed (must be cleared
;            in software)
;        0 = TMR0 register did not overflow
; bit 1: INTF: RB0/INT External Interrupt Flag bit
;        1 = The RB0/INT external interrupt occurred (must
;            be cleared in software)
;        0 = The RB0/INT external interrupt did not occur
; bit 0: RBIF: RB Port Change Interrupt Flag bit
;        1 = At least one of the RB7:RB4 pins changed
;            state (must be cleared in software)
;        0 = None of the RB7:RB4 pins have changed state
;
; T1CON is the Timer1 control register.
```

(*continued*)

Program 6.4 Timer1 usage. Rudimentary timer operation which depends on value of POT1 (*continued*)

```
T1CON      =%00000001
; bit 7-6: Unimplemented: Read as '0'
; bit 5-4: T1CKPS1:T1CKPS0: Timer1 Input Clock Pre-scale
;          Select bits
;          11 = 1:8 Pre-scale value
;          10 = 1:4 Pre-scale value
;          01 = 1:2 Pre-scale value
;          00 = 1:1 Pre-scale value
; bit 3:   T1OSCEN: Timer1 oscillator Enable Control bit
;          1 = oscillator is enabled
;          0 = oscillator is shut off (The osc inverter is
;              turned off to eliminate power drain)
; bit 2:   T1SYNC: Timer1 External Clock Input
;          Synchronization Control bit
;          TMR1CS = 1
;          1 = Do not synchronize external clock input
;          0 = Synchronize external clock input
;          TMR1CS = 0
;          This bit is ignored. Timer1 uses the internal
;          clock when TMR1CS = 0.
; bit 1:   TMR1CS: Timer1 Clock Source Select bit
;          1 = External clock from pin RC0/T1OSO/T1CKI (on
;              the rising edge)
;          0 = Internal clock (FOSC/4)
; bit 0:   TMR1ON: Timer1 On bit
;          1 = Enables Timer1
;          0 = Stops Timer1
;
; The option register
OPTION_REG = %00000000 ; Set Bit 7 to 0 and enable PORTB
                       ; pullups
                       ; All other bits are for Timer0 and
                       ; not applicable here
PIE1 = %00000001       ; See data sheet, enables interrupt.
ADCON0 = %11000001     ; Configure and turn on A/D Module
; bit 7-6: ADCS1:ADCS0: ; A/D Conversion Clock Select bits
;          00 = FOSC/2
;          01 = FOSC/8
;          10 = FOSC/32
;          11 = FRC (clock derived from an RC oscillation)
; bit 5-3: CHS2:CHS0: Analog Channel Select bits
;          000 = channel 0, (RA0/AN0)
;          001 = channel 1, (RA1/AN1)
;          010 = channel 2, (RA2/AN2)
```

(*continued*)

Program 6.4 Timer1 usage. Rudimentary timer operation which depends on value of POT1 (*continued*)

```
;           011 = channel 3, (RA3/AN3)
;           100 = channel 4, (RA5/AN4)
;           101 = channel 5, (RE0/AN5)(1)
;           110 = channel 6, (RE1/AN6)(1)
;           111 = channel 7, (RE2/AN7)(1)
; bit 2: GO/DONE: A/D Conversion Status bit
;           If ADON = 1 See bit 0
;           1 = A/D conversion in progress (setting this bit
;               starts the A/D conversion)
;           0 = A/D conversion not in progress (bit is auto
;               cleared by hardware when the
;           A/D conversion is complete)
; bit 1: Unimplemented: Read as 0;
; bit 0: ADON: A/D On bit
;           1 = A/D converter module is operating
;           0 = A/D converter module is shut off and consumes
;               no operating current
;
; The A to D control register for Port A is ADCON1
; ADCON1 = %00000010   ; Set part of PORTA analog
; The relevant table is on page 112 of the data sheet
; There are a number of choices which give us analog
; capabilities on PORTA.0
; and allow the voltage reference between Vdd and Vss. We
; have chosen 0010
; on the third line down in the table
;
; Next let us set up the port pin directions
TRISA = %11111111         ; Set PORTA to all input
TRISB = %11110000         ; Set up PORTB for keyboard reads
PORTB.0 = 0               ; set so we can read row 1 only for
                          ; now
                          ;
ON INTERRUPT GOTO TICKINT       ; Tells the program where to go
                                ; on interrupt
                                ;
                                ; Initialize display and write
                                ; to top line
LCDOUT $FE, 1, $FE, $80, "MM SS T" ;
                                   ;
MAINLOOP:                 ;
ADCON0.2 = 1              ; Conver'n to reads POT-1. Conver'n
                          ; start now and
```

(continued)

Program 6.4 Timer1 usage. Rudimentary timer operation which depends on value of POT1 (*continued*)

```
                                ; takes place during loop. If loop
                                ; was short we would
                                ; allow for that.
                                ; Then check the buttons to decide
                                ; what to do
IF PORTB.4 = 0 THEN STARTCLOCK      ;
IF PORTB.5 = 0 THEN STOPCLOCK       ;
IF PORTB.6 = 0 THEN CLEARCLOCK      ;
                                    ;
                                    ; and display what the
                                    ; clock status is
LCDOUT $FE, $80, DEC2 MINS, ":",DEC2 SECS, ":", DEC TENTHS,
" POT1=",DEC _
ADVAL, " "
                               ; We are now ready to read what
                               ; potentiometer setting is.
ADVAL = ADRESH                 ; we assumed that enough time
                               ; has passed to have an
                               ; updated value in the
                               ; registers. If not add wait
                               ; here.
GOTO MAINLOOP                  ; Do it again
                               ;
DISABLE                        ; Disable interrupts during
                               ; interrupt handler

TICKINT:                       ;
  TICKS = TICKS + 1            ; ticks are influenced by the
                               ; setting of POT-1
  IF TICKS < 5 THEN TIEXIT     ; arbitrary value to get one
                               ; second

  TICKS = 0                    ;
                               ;
  TENTHS = TENTHS + 1          ;
  IF TENTHS <9 THEN TIEXIT     ;
  TENTHS = 0                   ;
                               ;
  SECS = SECS + 1              ; update seconds
  IF SECS < 59 THEN TIEXIT     ;
  SECS = 0                     ;
  MINS = MINS + 1              ; update minutes
  TIEXIT:                      ;
  IF PORTB.5 = 0 THEN STOPCLOCK     ;
  TMR1H = ADRESH               ;
  PIR1 = 0                     ;
```

(*continued*)

Program 6.4 Timer1 usage. Rudimentary timer operation which depends on value of POT1 (*continued*)

```
RESUME                          ; go back to the main routine
ENABLE                          ;
                                ;
DISABLE                         ;
STARTCLOCK:                     ;
  INTCON = %10100011            ; Enable TMR1 interrupts
  TICKS = 0                     ;
GOTO MAINLOOP                   ;
                                ;
STOPCLOCK:                      ;
  INTCON = %10000011            ; Disable TMR1 interrupts
  PAUSE 2                       ;
  TICKS = 0                     ;
GOTO MAINLOOP                   ;
                                ;
CLEARCLOCK:                     ;
  INTCON = %10000011            ; Disable TMR1 interrupts
  MINS = 0                      ;
  SECS = 0                      ;
  TENTHS = 0                    ;
  TICKS = 0                     ;
GOTO MAINLOOP                   ;
ENABLE                          ;
END                             ;
```

Run this program to see how the setting of the potentiometer affects the operation of the stopwatch. It becomes clear that choosing how the interrupt will serve our purposes is very important, and a bad choice can compromise the operation of the program.

We can read the timer and the interrupts at our discretion either before or after an interrupt has occurred, and the interrupt flag can be cleared whenever we want to, if it has been set. If it has not been set there is no need to clear it. A simplified flow diagram is provided in Figure 6.1 to help you to understand what is going on in an interrupt servicing routine.

Even the 16-bit Timer1 on the 16F877A cannot time a period of any length. Repeated intervals have to be put together to create long time periods. The longest possible time between interrupts for Timer1 (with a 4 MHz clock) is 0.524288 seconds. The maximum pre-scale value is 1:8. The post-scaler is only available on Timer2, which in any case is an 8-bit timer. This results in a maximum time that is determined by multiplying the instruction clock cycle (1 µS @ 4 MHz) by the pre-scale (8) by the number of counts from one overflow to the next (65536):

```
1 µS * 8 * 65536 = 0.524288 seconds
```

On a 20 MHz machine the time would be one-fifth of this.

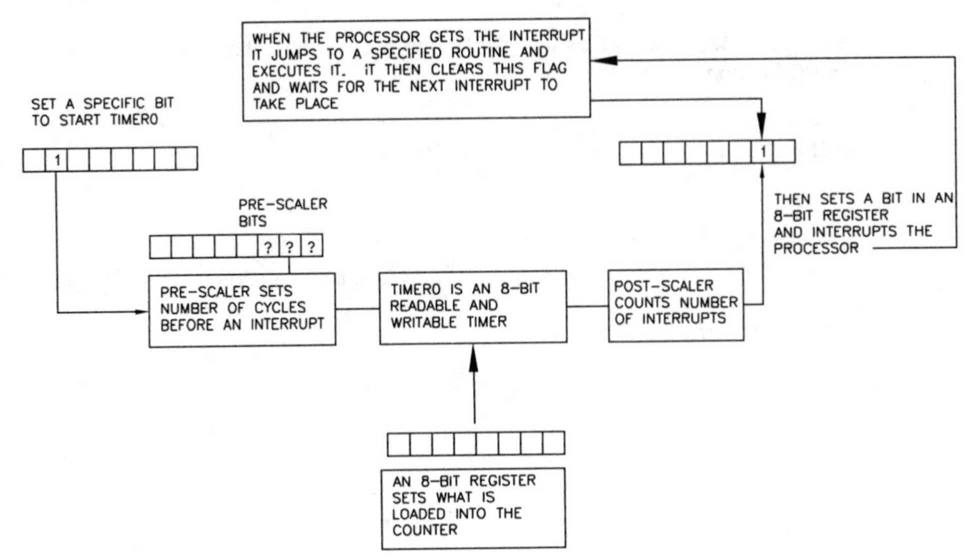

Figure 6.1 The simplified, basic flow diagram for a typical interrupt routine (Bits shown as being set in are not the real bits.)

Timer1, the 16-bit timer/counter, uses two registers: TMR1H and TMR1L. The timer has the following general properties:

- It increments from $0000 to $FFFF in two registers.
- If the interrupt is enabled, an overflow will occur when the 2-byte counter overflows from $FFFF to $0000.
- The device can be used as a timer.
- The device can be used as a counter.
- The timer registers can be read and written to.
- There is no post-scaler for this timer.

Simply stated again, this timer is used by setting its register to a selected value and using the interrupts this value creates for our purpose. A 16-bit timer will count up from where set to 65,535 and then flip to the selected value and start over again. An interrupt occurs and the interrupt flag is set every time the register overflows from 65,535 to 0. We do whatever needs to be done in response to the interrupt, resetting the interrupt flag and then going back to the main routine. On timers that permit the use of a pre-scaler and post-scaler, the pre-/post-scaler allows us to increase the time between interrupts by multiplying the time between interrupts with a definable value in a three to eight bit location. On writable timers we have the ability to start the timers with values of our choice in the timer register(s). This gives us very usable but not absolute control over the interrupt intervals.

Consider the fact that a 0.01 second timer setting with a pre-scaler set to 16 would provide us with an interrupt every 0.16 seconds. We would have 0.16 seconds to do whatever we wanted to do between interrupts. There are limits to what can be put in the timer counter and what can be put in the pre-scaler, and the interrupt frequency is also affected by the accuracy of the processor clock oscillator.

The value of the scaling that will be applied to the timer is determined by the contents of two bits in the interrupt control register. These bits multiply the time between interrupts by powers of 2 as under:

Pre-scaler	For Timer1	For the Watchdog Timer	
00	Multiply by 1	00	Multiply by 2
01	Multiply by 2	01	Multiply by 4
10	Multiply by 4	10	Multiply by 8
11	Multiply by 8	11	Multiply by 16

The two bits in the preceding table are bits 4 and bit 5 of the Timer1 control register T1CON.

The eight bits in T1CON are assigned as follows:

TMR1ON	Bit 0	1=Enables Timer1	0=Disables timer
TMR1CS	Bit 1	1=Use external clock	0=Use internal clock
TISYNC	Bit 2	1=Sync with internal clock input	0=Sync with external clock input
TIOSCEN	Bit 3	1=Enable oscillator	0=Shut off oscillator
T1CKPS1	Bit 4	Counter scaler is described in the pre-scaler table	
T1CKPS0	Bit 5	Counter scaler is described in the pre-scaler table	
	Bit 6	Not used	
	Bit 7	Not used	

Using Timer1 to Run a Critical Interrupt Driven Task While the Main Program Runs a Foreground Task

Let us use this timer in the same way we used Timer0 earlier and see what the differences between the two timers are. Since Timer1 is 16 bits wide, it can take much longer for it to set its interrupt flag. The interrupt flag was set approximately 61 times a second by Timer0. The Timer1 flag can take approximately 0.524 seconds, as calculated earlier. Let us write a short Timer1 program that is similar to the original Timer0 blinker program to see what the differences are.

Program 6.5 blinks the LEDs at D0 and D1 on and off alternately as the foreground part of the program. The interrupts generated by Timer1 are used to blink D3 on and off at half-second intervals. Since the control of D3 is driven by the interrupt, the timing stays accurate. Any time used by the interrupt routine is lost by the foreground task and affects the overall frequency of D0/D1 blink rate.

Program 6.5 Using TIMER1. Program blinks two LEDs alternately and blinks a third LED approx. half second on and half second off.

```
CLEAR                         ; clear memory
DEFINE OSC 4                  ; osc speed
TRISD = %00000000             ; set all PORTD lines to output
TRISE = %00000000             ; set all PORTE lines to output
                              ; set the A to D control register for
                              ; digital ports D, E
ADCON1 = %00000111            ; for 16F877A because it has analog
                              ; properties
T1CON = %00000001             ; turn on Timer0, pre-scaler = 1
INTCON = %11000000            ; enable global interrupts,
                              ; peripheral interrupts
                              ;
I VAR WORD                    ; counter variable
J VAR WORD                    ; counter variable
PAUSE 500                     ;
I = 0                         ; set counters to 0
J = 0                         ;
PIE1.0 = 1                    ; enable TMR1 overflow interrupt
ON INTERRUPT GOTO INTHANDLER ;
PORTD = 0                     ; turn off the entire port
                              ;
MAINLOOP:                     ;
IF PORTD.1 = 0 THEN           ; routine lights D0 and D1
                              ; alternately to
  PORTD.1 = 1                 ; that the program is running the
                              ; main routine
  PORTD.0 = 0                 ;
ELSE                          ;
  PORTD.1 = 0                 ;
  PORTD.0 = 1                 ;
ENDIF                         ;
FOR I = 1 TO 300              ; this is in lieu of a long pause
                              ; instruction
  PAUSEUS 100                 ; so that an interrupt is not
                              ; compromised
NEXT I                        ;
GOTO MAINLOOP                 ; do it all forever
                              ;
DISABLE                       ;
INTHANDLER:                   ; this is the interrupt service
                              ; routine
IF J < 6 THEN                 ; this routine allows 6
  J = J+1                     ; interrupts for each change of state
  GOTO COUNTNOTFULL           ; of LED D3.
```

(continued)

Program 6.5 Using TIMER1. Program blinks two LEDs alternately and blinks a third LED approx. half second on and half second off. (*continued*)

```
ELSE                      ;
  J = 0                   ;
ENDIF                     ;
IF PORTD.3 = 1 THEN       ; the D3 blink routine
  PORTD.3 = 0             ;
ELSE                      ;
  PORTD.3 = 1             ;
ENDIF                     ;
COUNTNOTFULL:             ;
PIR1.0 = 0                ; must now clear the interrupt flag
RESUME                    ;
ENABLE                    ;
END                       ; end program
```

Play with the value of the counter J to see how this affects the operation of the program. Study the differences between the two programs to set and clear the timer flags. Though Programs 6.4 and 6.5 do essentially the same thing, the setting of the potentiometer in the first program has to be modified to match the needs of the timer being used.

TIMER2: THE THIRD TIMER

Timer2 is an 8-bit timer only. It cannot be used as a counter. It has a pre-scaler and a post-scaler. The timer register for this counter is both writable and readable. If you can write to a counting register, you can set the value the count starts at and thus control the interval between interrupts (to some degree). That and the ability to set the pre- and post-scalers gives you the control you need for effective control of the interrupt interval, although you still cannot time all events exactly because of the coarseness of the settings available. Timer2 has a period register, PR2, which can be set by the user. The timer counts up from $00 to the value set in PR2, and when the two are the same it resets to $00. Small values in PR2 can be used to create very rapid interrupts—so much so that there may be no time to do anything else.

The Timer2 control register is T2CON. Its 8 bits are assigned as follows:

T2CKPS0	Bit 0	Counter pre-scaler	0 = Disables timer
T2CKPS1	Bit 1	Counter pre-scaler	0 = Uses internal clock
TMR2ON	Bit 2	1 = Timer2 is on	0 = Timer2 is off; shuts off oscillator
TOUTPS0	Bit 3	) Counter post-scaler value	
TOUTPS1	Bit 4	) Counter post-scaler value	
TOUTPS2	Bit 5	) Counter post-scaler value	
TOUTPS3	Bit 6	) Counter post-scaler value	
	Bit 7	Not used	

As always, the input clock for this timer is divided by 4 before it is fed to the timer. On a processor running at 4 MHz, the feed to the timer is at 1 MHz.

Pre-scaler for Timer2		Post-scaler for Timer2	
00	Multiply by 1, no scaling	0000	Multiply by 1, no scaling.
01	Multiply by 4	0001	Multiply by 2
1x	Multiply by 16	0010	Multiply by 3
		0011	Multiply by 4
		0100	Multiply by 5
		0101	Multiply by 6
		0110	Multiply by 7
		0111	Multiply by 8
		1000	Multiply by 9
		1001	Multiply by 10
		1010	Multiply by 11
		1011	Multiply by 12
		1100	Multiply by 13
		1101	Multiply by 14
		1110	Multiply by 15
		1111	Multiply by 16

The timer is turned on by setting bit 2 in register T2CON (the Timer2 control register).

The interrupt for Timer2 is enabled by setting Bit 1 of PIE1, and the system lets the program know that an interrupt has occurred by setting Bit 1 in PIR1. (Bit 0 in both these registers is for Timer1.)

Bit 7 (the global interrupt enable bit) of INTCON (the interrupt control register) enables all interrupts, including those created by Timer2. Bit 6 of INTCON enables all unmasked peripheral interrupts and is one of the ways to awaken a sleeping MCU.

Timer2 can also control the operation of the two PWM signals that can be programmed to appear on lines C1 and C2 with the HPWM command in PICBASIC PRO. Since one timer must control both lines, they both have to have the same PWM frequency. The relative width of the pulse within each of the PWM signals during each cycle does not have to be the same.

Timer2 is also used as a baud rate clock timer for communications. See Chapter 7 of the data sheet.

The Watchdog Timer

A Watchdog Timer is a timer that sets an interrupt that tells us that for some reason the program has hung up or otherwise gone awry. As such it is expected that in a properly written and compiled program the Watchdog Timer will never set an interrupt. This is accomplished by resetting the Watchdog Timer every so often in the program. The compiler inserts these instructions automatically into the program if the Watchdog Timer option is selected. However, setting the option does not guarantee that a program cannot or will not hang up. Software errors and infinite loops that reset the timer within themselves can still cause hang ups.

The Watchdog Timer is scalable. It shares its scaler with Timer0 on an exclusive basis. Either it uses the scaler, or Timer0 uses it. They cannot both use it at the same time. See the discussion under Timer0 in the data sheet for more information (Chapter 5).

Since PICBASIC PRO assumes that the Watchdog Timer will be run with a 1:128 pre-scaler, unwanted Watchdog Timer resets could occur when you assign the pre-scaler to Timer0. If you change the pre-scaler setting in OPTION_REG, you should disable the Watchdog Timer when programming. The Watchdog Timer enable/disable option is found on the configuration screen of your (hardware) programmer's software.

Counters

Of the three timers in the 16F877A, only Timer0 (an 8-bit timer) and Timer1 (the 16-bit timer) can be used as counters. Timer2 does not have a counter input line. Generally speaking this makes Timer0 suitable for use with small counts and rapid interrupts and Timer1 suitable for larger counts.

HOW DOES A COUNTER WORK?

The operation of a counter is similar to the operation of a timer except that instead of getting its count from an internal clock or oscillator, the counter gets its signals from an outside source. This means we have to do the following to use a counter:

- Decide which counter (timer) to use.
- Tell the counter where the signal is coming from.
- Tell it whether to count on a rising or falling edge of the signal.
- Decide what target count we are looking for, if applicable.
- Tell the counter where to start counting (because the interrupt will occur when the counter gets full).
- Decide whether we will need to scale the count by setting the scaler(s).

Once you start a counter, the counting continues until you deactivate it. There is no way to stop it or any purpose in doing so. It will reset if the MCU is reset, and it can be reset by writing to it. The rest has to do with knowing what bits to set in the control registers to get the counters to operate in the way we want them to.

USING TIMER0 AS A COUNTER

The three registers used to control the operation of the Timer0 module are TMR0, INTCON, and OPTION_REG. The interrupt control register is INTCON. See the data sheet.

Note *Though often called Timer0 and referred to as Timer0 here and in the data sheet, the real designation of this timer address is TMR0.*

This counter has the following properties:

- 8-bit timer/counter.
- Readable and writable.
- 8-bit software programmable pre-scaler.
- Internal or external clock select.
- Interrupt on overflow from FF (hex) to 00 (hex).
- Edge selection available for external clock signal.
- Counter mode is selected by setting OPTION_REG.5=1.
- External input for the timer must come in on PORTA.4 (pin 6 on the PIC).
- The edge direction is selected in OPTION_REG.4 (1=Rising edge).
- The pre-scaler is assigned with bits 0 to 3 of the OPTION_REG register (Chapter 5).

Note *The Watchdog Timer cannot use the pre-scaler when the pre-scaler is being used by Timer0.*

Since this is an 8-bit counter, it is suited to the counting a small number of counts, but longer counts can be accommodated by creating a routine to keep track of the number of interrupts.

Let us use the counter to count the pulses received from a 42-slot encoder mounted on a small DC motor. This same source will be used later for the Timer1 experiment for comparison between counters. (This motor will also be used later on as the encoded servomotor for the servo control experiments.)

We will set up the LCD display so that we can display certain registers during the operation of the program. We will also set the program up to read the potentiometers so that we can use their values to modify the program as the motor runs. Only POT0 and POT1 are used in this program.

Program 6.6 **Using Timer0 as a counter.** Counts the pulses from a motor driven encoder

```
CLEAR                          ; clear memory
DEFINE OSC 4                   ; 4 MHz clock
DEFINE LCD_DREG PORTD          ; data register
```

(continued)

Program 6.6 Using Timer0 as a counter. Counts the pulses from a motor driven encoder (*continued*)

```
DEFINE LCD_RSREG PORTE           ; register select
DEFINE LCD_RSBIT 0               ; pin number
DEFINE LCD_EREG PORTE            ; enable register
DEFINE LCD_EBIT 1                ; enable bit
DEFINE LCD_RWREG PORTE           ; read/write register
DEFINE LCD_RWBIT 2               ; read/write bit
DEFINE LCD_BITS 8                ; width of data
DEFINE LCD_LINES 2               ; lines in display
DEFINE LCD_COMMANDUS 2000        ; delay in µs
DEFINE LCD_DATAUS 50             ; delay in µs
                                 ;
DEFINE CCP1_REG PORTC            ; define the hpwm settings
DEFINE CCP1_BIT 2                ;
                                 ; define the A2D values
DEFINE ADC_BITS 8                ; set number of bits in result
DEFINE ADC_CLOCK 3               ; set internal clock source
                                 ; (3=rc)
DEFINE ADC_SAMPLEUS 50           ; set sampling time in µs
                                 ; set the analog to digital
                                 ; control register
ADCON1 = %00000110               ; needed for the 16F877A LCD
TEST VAR WORD                    ; assign the variable to be
                                 ; used
ADVAL0 VAR BYTE                  ; create adval to store result
ADVAL1 VAR BYTE                  ; create adval to store result
X VAR WORD                       ;
Y VAR WORD                       ;
PAUSE 500                        ; LCD start up
LCDOUT $FE, 1                    ; clear display
OPTION_REG = %00110000           ;
TMR0 = 0                         ; set up the register i/o
TRISC = %11110001                ; PORTC.0 is going to be the
                                 ; input to
                                 ; start the motor in that we
                                 ; are using a motor
                                 ; encoder for input
PORTC.3 = 0                      ; enable the motor
PORTC.2 = 0                      ; set the rotation direction
                                 ;
LOOP:                            ;
ADCIN 0, ADVAL0                  ; read channel 0 to ADVAL0
ADCIN 1, ADVAL1                  ; read channel 1 to ADVAL1
ADCIN 3, ADVAL2                  ; read channel 3 to ADVAL2
                                 ;
```

(continued)

Program 6.6 Using Timer0 as a counter. Counts the pulses from a motor driven encoder (*continued*)

```
TMR0 = 0                       ;
PAUSE ADVAL1                   ;
X = TMR0                       ;
IF ADVAL0>20 THEN              ;
 HPWM 2, ADVAL0, 32000         ;
 LCDOUT $FE, $80, DEC4 X," ",DEC ADVAL1," "      ;
 LCDOUT $FE, $C0, "PWM = ",DEC ADVAL0," " ;
ELSE                           ;
 LCDOUT $FE, $C0, "PWM TOO LOW ",DEC ADVAL0," " ;
ENDIF                          ;
                               ;
GOTO LOOP                      ;
END                            ;
```

You can change the speed of the motor and the time interval for the counts with the two potentiometers.

Play with the values of the two potentiometers to see what happens. Be careful about overflowing the counters past 255.

The Timer0 counter is affected by the OPTION REGISTER bits as follows:

```
OPTION_REG.6 = 1     ; Interrupt on rising edge
OPTION_REG.5 = 0     ; External clock
OPTION_REG.4 = 1     ; Increment on falling edge not used
OPTION_REG.3 = 0     ; Assign pre-scaler to Timer0
OPTION_REG.2 = 1     ; ] These three bits set the pre-scaler
OPTION_REG.1 = 1     ; ] You can experiment with changing these three
OPTION_REG.0 = 1     ; ] bits to see what happens
```

Put these and other values in the program and run the program. See what happens.

USING TIMER1 AS A COUNTER

The operation of Timer1 as a counter is similar to the operation of Timer0. However, because Timer1 is a 16-bit timer, much longer counts can be handled and counts coming in at faster rates can be counted. It also means that a lot more can be done in the TimerLoop and BlinkerLoop routines if the program is designed to do so. However, the set up for Timer1 is more complicated because of the more numerous options available.

The differences between the use of the two timers have to do with the setup of the controlling registers. Timer1 is controlled by six registers, compared to the three for Timer0. Timer1's registers are as follows:

INTCON — Interrupt control register
PIR1 — Peripheral interrupt register 1
PIE1 — Peripheral interrupt enable register 1
TMR1L — Low byte of the timer register
TMR1H — High byte of the timer register
T1CON — Timer1 interrupt control

Again, and as always, the frequency of the oscillator is divided by 4 before being fed to the counter when you use the internal clock.

Page 52 of data sheet describes how Timer0 is used with an external clock.

Counter mode is selected by setting bit TMR1CS. In this mode, the timer increments on every rising edge of clock input on pin RC1/T1OSI/CCP2, when bit T1OSCEN is set, or on pin RC0/T1OSO/T1CKI, when bit T1OSCEN is cleared.

Three of the pins on the 16F877A can be used as inputs to the Timer1 counter module:

- Pin PORTA.4, the external clock input (pin 6 on the PIC)
- Pin PORTC.0, selected by setting TIOSCEN =1
- Pin PORTC.1, selected by setting TIOSCEN=0

Timer1 is enabled by setting T1CON.0=1. It stops when this bit is turned off or disabled.

The clock that Timer1 will use is selected by T1CON.1. The external clock is selected by setting this to 1. The input for this external clock must be on PORTA.4.

In summary, eight bits in Timer1 control register T1CON, which provides the following functions:

- ~~**Bit 7**: Not used and read as a 0~~
- ~~**Bit 6**: Not used and read as a 0~~
- **Bit 5**: Input pre-scaler
- **Bit 4**: Input pre-scaler
- **Bit 3**: Timer1 oscillator enable
- **Bit 2**: Timer1 external clock synchronization
- **Bit 1**: Timer1 clock select
- **Bit 0**: Timer1 enable

If the interrupts are not going to be used all the other registers can be ignored, and we can set:

```
T1CON to %00110001
```

The setting of these bits is described in detail in Chapter 5 of the data sheet.

Program 6.7 reflects this discussion.

First let us define all the defines that we will need. Here all the defines are included as an example, but not all are needed when using the LAB-X1.

Program 6.7 **Timer1 as a counter.** Timer1 counts signals from a servomotor encoder.

```
CLEAR                               ; clear memory
DEFINE OSC 4                        ; 4 MHz clock
DEFINE LCD_DREG PORTD               ; data register
DEFINE LCD_RSREG PORTE              ; register select
DEFINE LCD_RSBIT 0                  ; pin number
DEFINE LCD_EREG PORTE               ; enable register
DEFINE LCD_EBIT 1                   ; enable bit
DEFINE LCD_RWREG PORTE              ; read/write register
DEFINE LCD_RWBIT 2                  ; read/write bit
DEFINE LCD_BITS 8                   ; width of data
DEFINE LCD_LINES 2                  ; lines in display
DEFINE LCD_COMMANDUS 2000           ; delay in micro seconds
DEFINE LCD_DATAUS 50                ; delay in micro seconds
```

The next two lines define which pin is going to be used for the HPWM signal that will control the speed of the motor. The encoder that we are looking at is attached to the motor.

```
DEFINE CCP1 REG PORTC               ; define the HPWM settings
DEFINE CCP1_BIT 2                   ; pin C1
```

The next three lines define the reading of the three potentiometers on the board. Only the first potentiometer is being used in the program, but the others are defined so that you can use them when you modify the program. The potentiometers give you values you can change in real time.

```
                           ; define the A2D values
DEFINE ADC_BITS 8          ; set number of bits in result
DEFINE ADC_CLOCK 3         ; set internal clock source (3=rc)
DEFINE ADC_SAMPLEUS 50     ; set sampling time in µS
```

Next, we set ADCON1 to bring the MCU back into digital mode. Since this PIC has analog capability, it comes up in analog mode after a reset or on startup.

```
                           ; set the Analog to Digital control
                           ; register
ADCON1 = %00000111         ; needed for the LCD operation
                           ; we create the variables that we
                           ; will need.
TMR1 VAR WORD              ; set the variable for the timer
ADVAL0 VAR BYTE            ; create adval to store result
ADVAL1 VAR BYTE            ; create adval to store result
```

(continued)

Program 6.7 Timer1 as a counter. Timer1 counts signals from a servomotor encoder. (*continued*)

```
ADVAL2 VAR BYTE          ; create adval to store result
X VAR WORD               ; spare variable for experimentation
Y VAR WORD               ; spare variable for experimentation
PAUSE 500                ; pause for LCD to start up
LCDOUT $FE, 1            ; clear Display and cursor home
                         ;
                         ; set up the register I/O
TRISC = %11110001        ; PORTC.0 is going to be the Input
CCP1CON = %00000101      ; capture every rising edge
T1CON = %00000011        ; no pre-scale/osc off/Sync on
                         ; external source/TMR1 on
                         ; start the motor, using a motor
                         ; encoder for input
PORTC.3 = 0              ; enable the motor
PORTC.2 = 1              ; set the rotation direction
```

Next, we will go into the body of the program. The loop starts with reading all three potentiometers, though we are using only the first one to set the power and thus the speed of the motor.

```
LOOP:                    ;
ADCIN 0, ADVAL0          ; read channel 0 to ADVAL0
ADCIN 1, ADVAL1          ; read channel 1 to ADVAL1
ADCIN 3, ADVAL2          ; read channel 3 to ADVAL2
```

If the duty cycle of the motor is less than 20 out of 255, the motor will not come on, so we make an allowance for that and display the condition on the LCD.

```
IF ADVAL0>20 THEN              ;
  HPWM 2, ADVAL0, 32000        ;
  LCDOUT $FE, $C0, "PWM = ",DEC ADVAL0," "        ;
ELSE       ;
  LCDOUT $FE, $C0, "PWM TOO LOW ",DEC ADVAL0," ";
ENDIF      ;
```

Then we read the two timer registers to see how may counts went by. In this case, the counts were too low to show up in the high bits, so the high bits were ignored. However, if you have a faster count input, you might want to add this information to the readout.

```
TMR1H = 0                      ; clear Timer1 high 8-bits
TMR1L = 0                      ; clear Timer1 low 8-bits
T1CON.0 = 1                    ; start 16-bit timer
PAUSE 100                      ; capture 100 ms of Input
                               ; Clock Frequency
```

(*continued*)

Program 6.7 **Timer1 as a counter.** Timer1 counts signals from a servomotor encoder. (*continued*)

```
TlCON.0 = 0                         ; stop 16-bit Timer
TMR1.BYTE0 = TMR1L                  ; read Low 8-bits
TMR1.BYTE1 = TMR1H                  ; read High 8-bits
TMR1 = TMR1 - 11                    ; capture Correction
IF TMR1 = 65525 THEN NOSIGNAL ; see PicBasic book for
                                    ; explanation.
  LCDOUT $FE, $80, DEC5 TMR1," COUNTS"     ; frequency
                                           ; display
  PAUSE 10                          ; slow down
GOTO LOOP                           ; do it again
                                    ;
NOSIGNAL:                           ;
  LCDOUT $FE, $80, "NO SIGNAL "     ;
GOTO LOOP                           ;
END                                 ;
```

Pre-scalers and Post-scalers

Pre-scalers and post-scalers can be confusing for the beginner. Here is a simple explanation.

A pre-scaler is applied to the system clock and affects the timer by slowing down the system clock as it applies to the timer. Normally the timer is fed by a fourth of the basic clock frequency, which is called Fosc/4. In a system running a 4 MHz, the timer sees a clock running at 1 MHz. If the pre-scaler is set for I:8, the clock will be slowed down by another eight times, and the timer will see a clock at 125 kHz. See Figure 6.2 (for Timer1) in the data sheet to see how this applies to Timer1.

A post-scaler is applied after the timer count exceeds its maximum value, generating an overflow condition. The post-scaler setting determines how may overflows will go by before an interrupt is triggered. If the post-scaler is set for 1:16, the timer will overflow 16 times before an interrupt flag is set. The upper diagram on page 55 (for Timer2) of the data sheet shows this in its diagrammatic form and is worth studying.

All other things being equal, both scalers are used to increase the time between interrupts.

When starting out, just leave the scalers at 1:1 values and nothing will be affected. We will not need them for any of the experiments that we will be doing right away. Once you learn the more sophisticated uses of timers you can play with the values and learn more about how to use them. The primary use is in creating accurate timing intervals for communications and so on, because no external routines are necessary when this is done with scalers. Everything becomes internal to the PIC and is therefore not affected by external disturbances.

Note *Additional information on timer modules is available in* The PICmicro Mid-Range MCU Family Reference Book *(DS33023).*

Timer Operation Confirmation

To make sure a timer is working, set up a program in which the interrupt routine increments a variable and the main loop displays it. If you see the variable incrementing, the interrupt routine is working.

The speed of the incrementing will give you some idea of the rate at which the interrupts are occurring and will confirm that the interrupts are occurring as fast as you have programmed them to.

Other timer related registers can also be looked at to see what is going on in them by displaying them on the LCD.

Caution *Writing to the LCD is time consuming and will slow down the system.*

Exercises for Timers

1. Write a program to generate a one minute timer clock with a 0.1 second display and then do the following:
   - Check its accuracy with the time site on the Internet.
   - Make adjustments to make it accurate to within one second per hour and then per day. Can this be done? Which timer works best? Which timer is the easiest to use for such a task? What are the problems that you identified?
2. Write the preceding program for each of the other two timers.

Exercises for Counters

1. Design and make a tachometer for small model aircraft engines. Make the range between 5 rev per second to 50,000 rev per minute displayed on the LCD in real time.
2. Design and build a thermometer based on the changes in frequency exhibited by a 555 timer circuit being controlled by a thermistor. Calibrate the thermometer with a lookup table. If you are not familiar with the use of lookup tables, research on the Internet so you can understand how to use them. They are very useful devices at the level that we are working.

7

CLOCKS AND MEMORY: SOCKETS U3, U4, U5, U6, U7, AND U8

Sockets U3, U4 and U5: For Serial One Wire Memory Devices

This chapter describes the interaction of PIC microprocessors with serial, one wire, and memory devices. These devices are suitable for adding limited amounts of memory to the microprocessors with a minimal number of interfacing lines. Though they are all referred to as one wire devices, none of them can be interfaced with a microcontroller with just one wire. However, the data does flow back and forth on one wire. Other wires are needed for power and timing as shown in Figure 7.1.

Each PIC microcontroller comes with a certain amount of on-board memory. This memory is enough for most of applications that are created, but there are times when more memory is needed to get the job done. There are five empty 8-pin sockets on the LAB-X1. The three of these on the left are designed to allow us to experiment with three types of single wire memory ICs. The ICs don't need just one wire for full control, but the data does go back and forth on one wire.

Note *Each memory socket accepts only one type of memory device and only one of the ICs is allowed to be in place at any one time. This is because the lines are shared between the sockets and having more than one device plugged in can create conflicts.*

Depending on the type of memory you want to experiment with, one of the three schematics shown in Figure 7.1 can be used.

The interfaces that have been developed for the three types of one wire memory give you the choices you need for flexibility in board design and layout, but they also mandate that a single interface and protocol will not work for everything. The interfaces vary in speed, number of signal lines, and other important details.

Since the memories are all one-wire serial devices, their memory content can vary from 128 bytes to 4 kilobytes or more and still maintain the 8-pin interface.

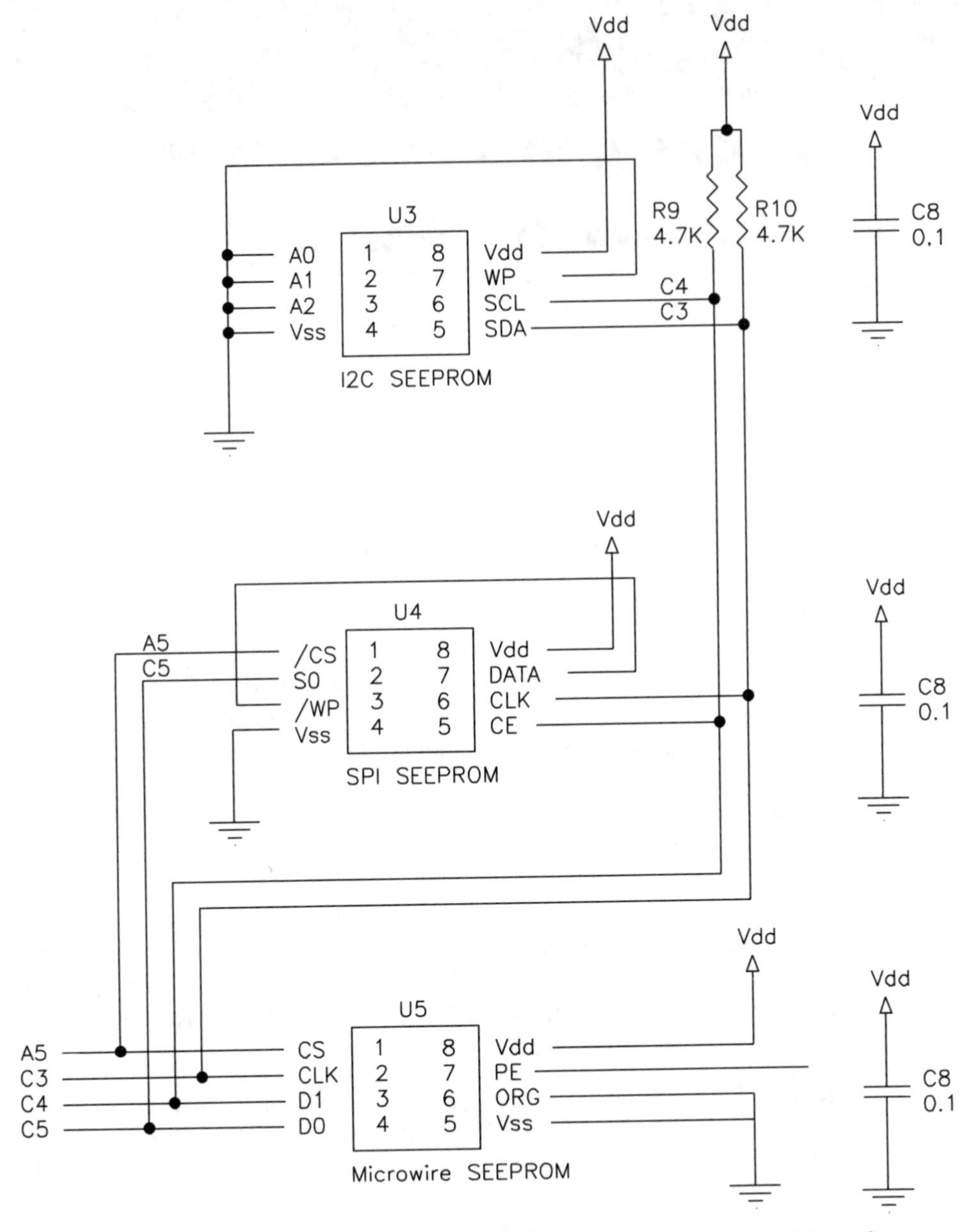

Figure 7.1 The three types of memory that you can experiment with on the LAB-X1 and the wiring layouts for each of them. Only one IC may be in place at any one time.

The salient characteristics of the three types of memory are as follows:

- **I2C SEEPROM** Serial electrically erasable and programmable read-only memories. These are best suited for applications needing a modest amount of inexpensive, nonvolatile memory where a lot of I/O lines are not available for memory transfers; requires 4 wires for control.
- **SPI** The Serial Peripheral Interface, originated at Motorola. SPI is much like Microwire, though the signal names, polarities, and other details vary. Like Microwire, SPI is often referred to as a 3-wire interface, though a read/write interface actually requires two data lines, a clock, a chip select, and a common ground, making five wires.
- **Microwire** A National Semiconductor standard Microwire is specially suited to use with their microcontrollers. Though often called a 3-wire interface, it too is actually a 5-wire interface.

Other manufacturers provide products to meet the wiring standards that are illustrated in Figure 7.1.

Which EEPROM type should you use? I2C is best if you have just two signal lines to spare, or if you have a cabled interface (I2C also has the strongest drivers). However, if you want a clock faster than 400 kilohertz, use Microwire or SPI.

For more on using serial EEPROMs, refer to the manufacturers' pages on the Web, especially these sites:

- **National Semiconductor** http://www.national.com/design/; many application notes on Microwire.
- **Motorola Semiconductor** http://www.mcu.motsps.com/mc.html; microcontroller references contain SPI documentation.

Jan Axelson's article in *Circuit Cellar Ink* magazine is a good source of detailed information on these devices. The article can be found at www.lvr.com/files/seeprom.pdf.

The PICBASIC PRO compiler provides all the instructions necessary to access these serial memories.

SOCKET U3: I2C SEEPROM

Socket U3 accommodates I2C memory only. The wiring arrangement needed to implement the use of these devices is shown in Figure 7.2.

Program 7.1, written by microEngineering Labs demonstrates one way or writing to and reading from I2C serial memory devices.

Program 7.1 Read from and write to I2C SEEPROMs

```
CLEAR                                   ; clear memory
SO CON 0                                ; define serial output pin
N2400 CON 4                             ; set serial mode
                                        ; define variables
DPIN VAR PORTA.0                        ; I2C data pin
```

(continued)

Program 7.1 **Read from and write to I2C SEEPROMs** (*continued*)

```
CPIN VAR PORTA.1                              ; I2C clock pin
B0   VAR BYTE                                 ;
B1   VAR BYTE                                 ;
B2   VAR BYTE                                 ;
                                              ; write to the memory
FOR B0 = 0 TO 15                              ; loop 16 times
  I2CWRITE DPIN, CPIN, $A0, B0, [B0]          ; write each
                                              ; location's address
                                              ; to itself
  PAUSE 10                                    ; delay 10 ms after
                                              ; each write is
                                              ; needed
NEXT B0                                       ;
                                              ;
LOOP:                                         ;
FOR B0 = 0 TO 15 STEP 2                       ; loop 8 times
  I2CREAD DPIN, CPIN, $A0, B0, [B1, B2]       ; read 2 locations in
                                              ; a row
  SEROUT SO, N2400, [#B1," ",#B2," "]         ; print 2 locations
                                              ; to CRT
NEXT B0                                       ;
                                              ;
SEROUT SO, N2400, [13,10]                     ; print linefeed
GOTO LOOP                                     ;
END                                           ;
```

SOCKET U4: SPI SEEPROM

Socket U4 is wired to use SPI memory only. The wiring arrangement needed to implement the use of these devices is shown in Figure 7.3.

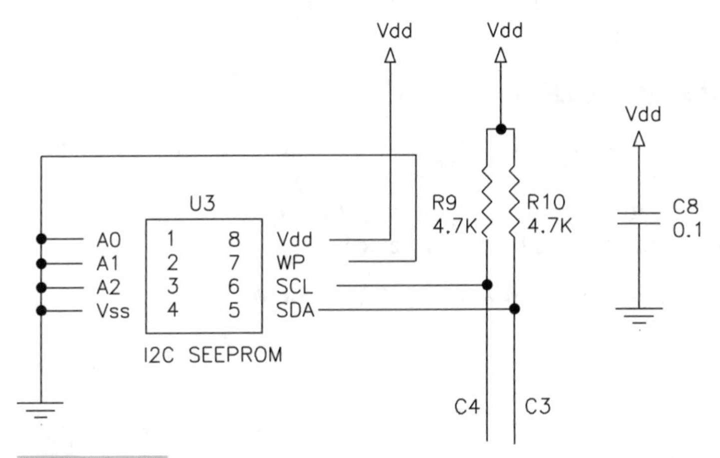

Figure 7.2 **12C memory: Wiring and circuitry requirements**

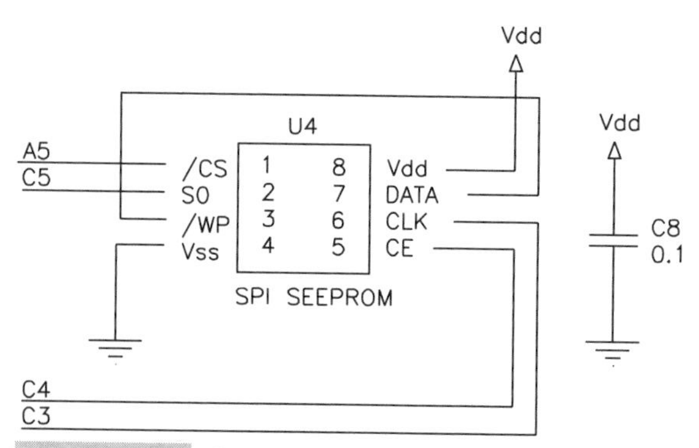

Figure 7.3 SPI SEEPROM: Wiring and circuitry requirements

Program 7.2 was written by microEngineering Labs. It demonstrates writing to and reading from SPI SEEPROM memory devices. It does this by first writing to the first 16 locations of an external serial EEPROM. It then reads these 16 locations back into the LAB-X1 and sends the data read to the LCD and repeats the process.

Program 7.2 Read from and write to SPI SEEPROMs.

```
DEFINE LOADER_USED 1              ; LOADER_USED to allow use of
                                  ; the boot loader.
                                  ; this will not affect normal
                                  ; program operation.
DEFINE LCD_DREG  PORTD            ; define LCD registers and bits
DEFINE LCD_DBIT  4                ;
DEFINE LCD_RSREG PORTE            ;
DEFINE LCD_RSBIT 0                ;
DEFINE LCD_EREG  PORTE            ;
DEFINE LCD_EBIT  1                ;
INCLUDE "MODEDEFS.BAS"            ;
CS VAR PORTA.5                    ; chip select pin
SCK VAR PORTC.3                   ; clock pin
SI VAR PORTC.4                    ; data in pin
SO VAR PORTC.5                    ; data out pin
ADDR VAR WORD                     ; address
B0 VAR BYTE                       ; data
TRISA.5 = 0                       ; set CS to output
ADCON1 = %00000111                ; set all of PORTA and PORTE
                                  ; to digital
LOW PORTE.2                       ; LCD R/W line low (W)
PAUSE 100                         ; wait for LCD to start up
FOR ADDR = 0 TO 15                ; loop 16 times
  B0 = ADDR + 100                 ; B0 is data for SEEPROM
```

(continued)

Program 7.2 **Read from and write to SPI SEEPROMs** *(continued)*.

```
  GOSUB EEWRITE                       ; write to SEEPROM
  PAUSE 10                            ; delay 10 ms after each
                                      ; write
NEXT ADDR                             ;
LOOP: FOR ADDR = 0 TO 15              ; loop 16 times
  GOSUB EEREAD                        ; read from SEEPROM
  LCDOUT $FE, 1, #ADDR,": ",#B0       ; display
  PAUSE 1000                          ;
  NEXT ADDR                           ;
GOTO LOOP                             ;
; Subroutine to read data from addr in serial EEPROM
EEREAD: CS = 0                        ; enable serial EEPROM
  SHIFTOUT SI, SCK, MSBFIRST, [$03, ADDR.BYTE1, ADDR.BYTE0]
                                      ; Sends read cmd and addr
  SHIFTIN SO, SCK, MSBPRE, [B0]       ; read data
  CS = 1                              ; disable
RETURN                                ;
; Subroutine to write data at addr in serial EEPROM
EEWRITE: CS = 0                       ; enable serial EEPROM
  SHIFTOUT SI, SCK, MSBFIRST, [$06]   ; send write enable
                                      ; command
  CS = 1                              ; disable to execute
                                      ; command
  CS = 0                              ; enable
  SHIFTOUT SI, SCK, MSBFIRST, [$02, ADDR.BYTE1, ADDR.BYTE0, B0]
                                      ; Sends address and data
  CS = 1                              ; disable
RETURN                                ;
END                                   ;
```

SOCKET U5: MICROWIRE DEVICES

Socket U5 is wired to use Microwire memory. The wiring arrangement needed to implement the use of these devices is shown in Figure 7.4.

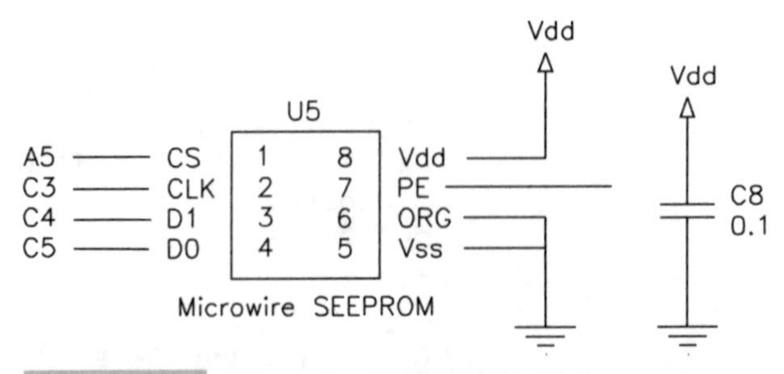

Figure 7.4 **Microwire SEEPROM: Wiring and circuitry requirements**

Program 7.3 was written by microEngineering Labs. The program reads from and writes to Microwire SEEPROM 93LC56A.

It writes to the first 16 locations of the external serial EEPROM. It then reads the 16 locations just written back into the LAB-X1 and sends them to the LCD for display. The process is repeated in a loop.

Note *The program is written for SEEPROMs with byte-sized address.*

Program 7.3 **Read from and write to Microwire SEEPROMs.**

```
DEFINE LCD_DREG PORTD           ; define LCD registers
                                ; and bits
DEFINE LCD_DBIT 4               ;
DEFINE LCD_RSREG PORTE          ;
DEFINE LCD_RSBIT 0              ;
DEFINE LCD_EREG PORTE           ;
DEFINE LCD_EBIT 1               ;
INCLUDE "MODEDEFS.BAS"          ;
CS          VAR    PORTA.5      ; chip select pin
CLK         VAR    PORTC.3      ; clock pin
DI          VAR    PORTC.4      ; data in pin
DO          VAR    PORTC.5      ; data out pin
ADDR VAR    BYTE                ; address
B0          VAR    BYTE         ; data
LOW CS                          ; chip select inactive
ADCON1 = 7                      ; set PORTA and PORTE to
                                ; digital
LOW PORTE.2                     ; LCD R/W line low (W)
PAUSE 100                       ; wait for LCD to start up
GOSUB EEWRITEEN                 ; enable SEEPROM writes
FOR ADDR = 0 TO 15              ; loop 16 times
  B0 = ADDR + 100               ; B0 is data for SEEPROM
  GOSUB EEWRITE                 ; write to SEEPROM
  PAUSE 10                      ; delay 10 ms after each write
NEXT ADDR                       ;
LOOP: FOR ADDR = 0 TO 15        ; loop 16 times
  GOSUB EEREAD                  ; read from SEEPROM
  LCDOUT $FE, 1, #ADDR,": ",#B0; display
  PAUSE 1000                    ;
  NEXT ADDR                     ;
GOTO LOOP                       ;
; Subroutine to read data from addr in serial EEPROM
EEREAD: CS = 1                  ; enable serial EEPROM
SHIFTOUT DI, CLK, MSBFIRST, [%1100\4, ADDR]; Send read
                                             ; command and
                                             ; address
SHIFTIN DO, CLK, MSBPOST, [B0]; read data
```

(continued)

Program 7.3 **Read from and write to Microwire SEEPROMs** (*continued*).

```
CS = 0                            ; disable
RETURN                            ;
                                  ; subroutine to write data at
                                  ; addr in serial EEPROM
EEWRITE: CS = 1                   ; enable serial EEPROM
SHIFTOUT DI, CLK, MSBFIRST, [%1010\4, ADDR, B0]
                                  ; sends write command, address
                                  ; and data
CS = 0                            ; disable
RETURN                            ;
; Subroutine to enable writes to serial EEPROM
EEWRITEEN: CS = 1                 ; enable SERIAL EEPROM
SHIFTOUT DI, CLK, MSBFIRST, [%10011\5, 0\7]
                                  ; sends write enable command
                                  ; and dummy clocks
CS = 0                            ; disable
RETURN                            ;
END
```

Socket U6: Real Time Clocks

There are four options for using socket U6. This socket is designed to let us experiment with three real time clocks and with a 12-bit analog-to-digital converter. The socket connects to the microcontroller, as shown in Figure 7.5.

As can be seen in Figure 7.5, there is a 5-wire plus ground interface between the MCU and the IC. The wiring for this chip is similar to the wiring for the Microwire SEEPROMS and the Microwire memory ICs. Essentially, this looks like a memory chip to the processor. When we write to this memory, we are writing to the clock; when we read from this chip, we are reading an ever-changing memory content that gives us information that we can interpret as "time." The program to read and write to this clock looks like a program that interacts with the Microwire family of SEEPROMS. The same is true for the other chips, as shown in Figures 7.6 and 7.7.

The NJU6355, the DS1202, and the DS1302 real time clocks are the three integrated circuits that can be used in socket U6. Note the following about these clocks:

- Jumper J5, which is used for soldering in the crystal for the clock ICs, is also the connection that the analog signal for the 12-bit A to D converter goes into. So if you solder in a crystal, you will have to remove the crystal and make arrangements to read in the analog signal when you want to experiment with the LTC1298 12-bit A to D converter. The A to D converter uses the same socket (U6) as is used by the three clock chips.
- There are a total of six empty sockets on the LAB-X1 board as received: U3, U4, U5, U6, U7, and U10. Though more than one socket can be occupied by an IC at any one time, it is best if only one IC is experimented with at any one time.

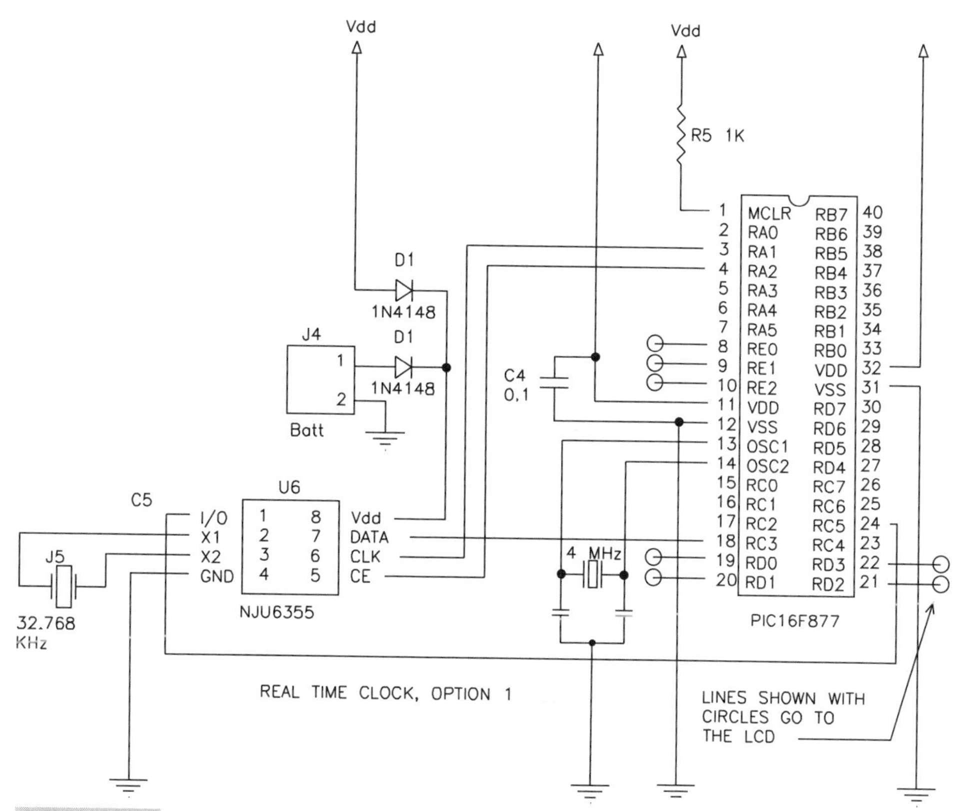

Figure 7.5 Clock implemented using IC NJU6355: this clock IC looks like a set of memory locations to the MCU.

This will ensure that there are no conflicts between the various devices. If the extreme right RS485 socket U10 is to be used, the RS 232 IC in the socket just to the left of it, in U9, has to be removed. One of these two communication chips can remain in place at all times and will not conflict with the memory locations.

THE CLOCK ICS IN SOCKET U6

The two 8-pin Dallas Semiconductor clock ICs are interchangeable, and each of them goes into socket U6. The DS1302 is the successor to the DS1202.

The NJU6355 also goes into socket U6, but it is not pin-for-pin compatible with the Dallas Semiconductor chips. Fortunately, it too needs to have its crystal between pins 2 and 3, and its other lines can share the connections to the PIC 16F877A.

Before you can use the NJU6355, the DS1202, or the DS1302, you have to install a crystal between pins 2 and 3 of the chip socket. This has to be a 32.768 KHz crystal, and it is to be installed at jumper J5 next to the real time clock IC socket. If you do not have a crystal in place, the program will show the date and other items on the LCD, but the clock and the date will not move forward.

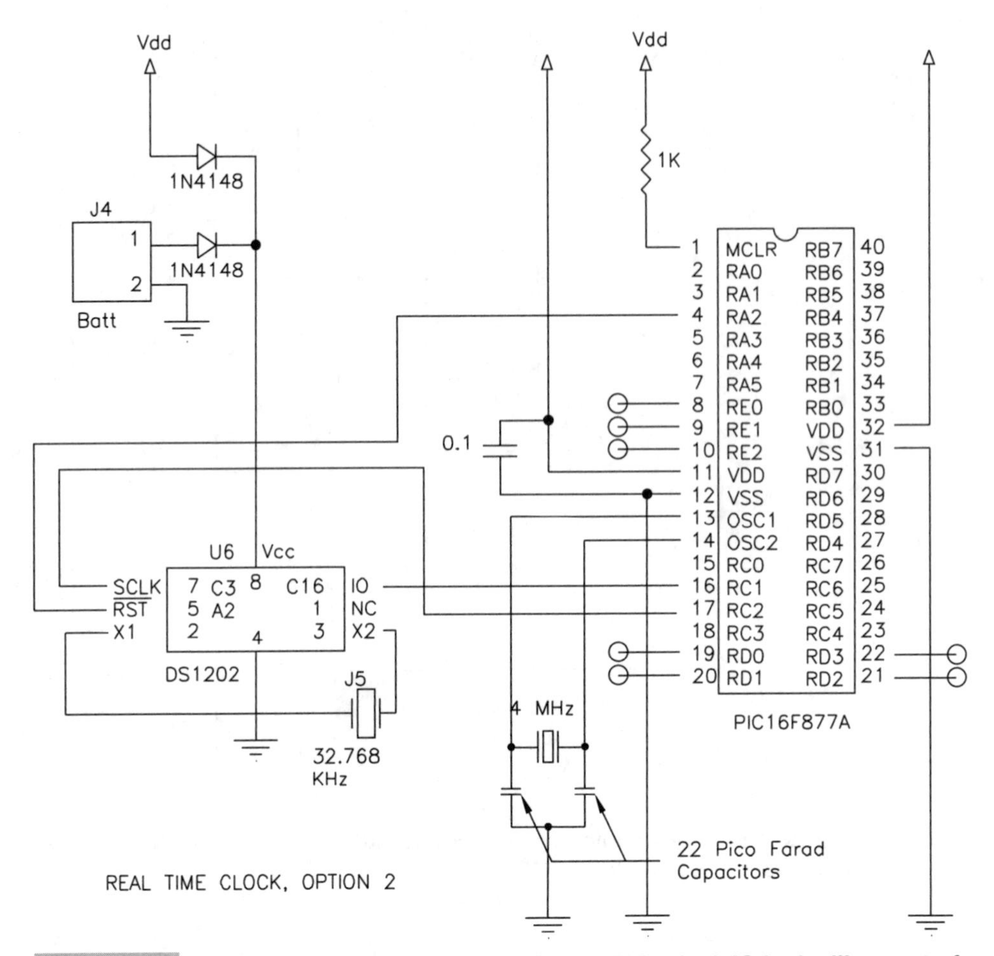

Figure 7.6 Clock implemented using IC DS1202: This clock IC looks like a set of memory locations to the MCU.

If you want to have battery back up for the clock, you need to install a battery at jumper J4 at the edge of the board next to U10. The pins for this jumper are already on the board when you receive it. The IC will accept from 2.0 to 5.5 V, so three AA cells in series can provide an inexpensive backup power source. (The power drawn by this IC is 300 nanoamps at 2 V. Two AA cells may not provide enough voltage because of the voltage drop across the in-line diode.)

THE DS1302 IN SOCKET U6

The DS1302 is the successor to the DS1202. They are very similar except for the DS1302's backup power capability and seven additional bytes of scratch pad memory. See the data sheet for more specific details.

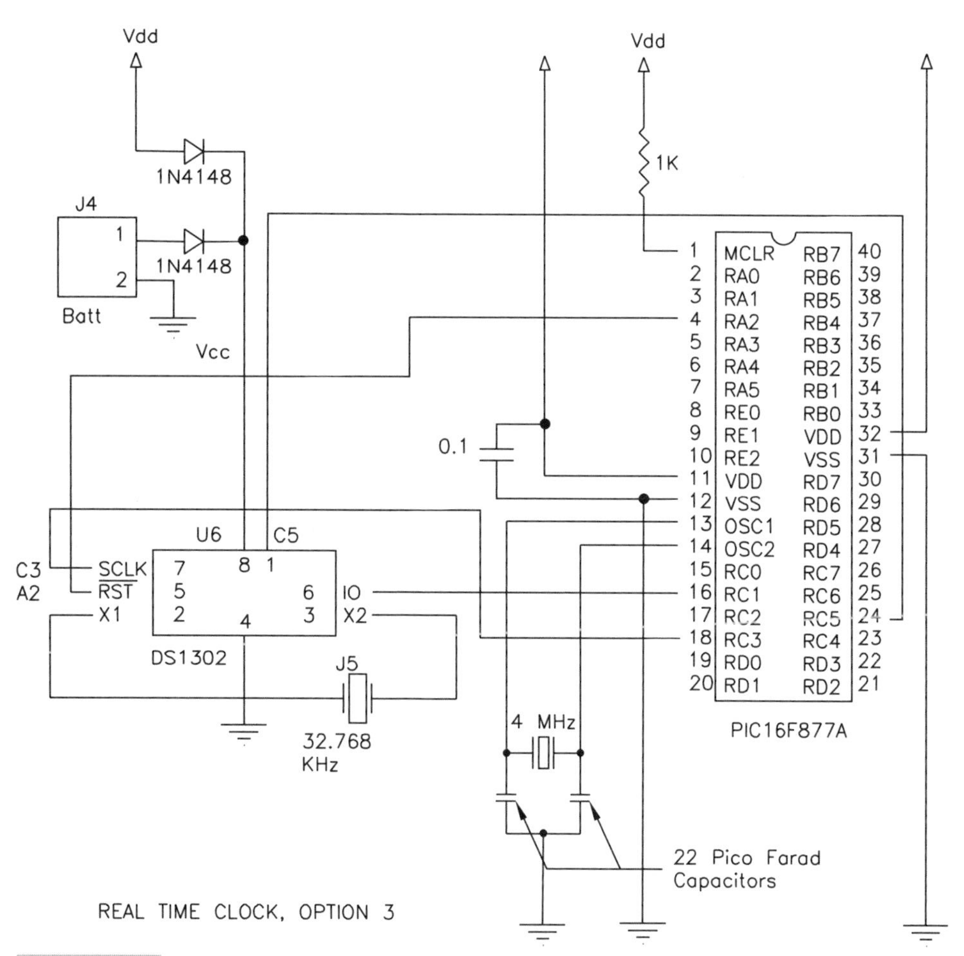

Figure 7.7 Clock implemented using IC DS1302: This clock IC looks like a set of memory locations to the MCU.

The emphasis in the program we will develop is to see how we get the data to and from the real time clock. Setting the clock is going to be done in the program startup routine, and the time cannot be modified once the program is running. If you want to modify it, you can add whatever is necessary to do to the program you write.

There are 31 RAM registers in the DS1302. When you want to send or receive data to the IC, the data can be transferred to and from the clock/RAM one byte at a time or in a burst of up to 31 bytes.

THE LTC1298 12-BIT A TO D CONVERTER IN SOCKET U6

For our purposes 8-, 10-, and 12-bit A to D converters are used as interfaces between sensors and microprocessors. Sensors usually provide a change in resistance, inductance, or capacitance as some other factor is manipulated. These changes are usually

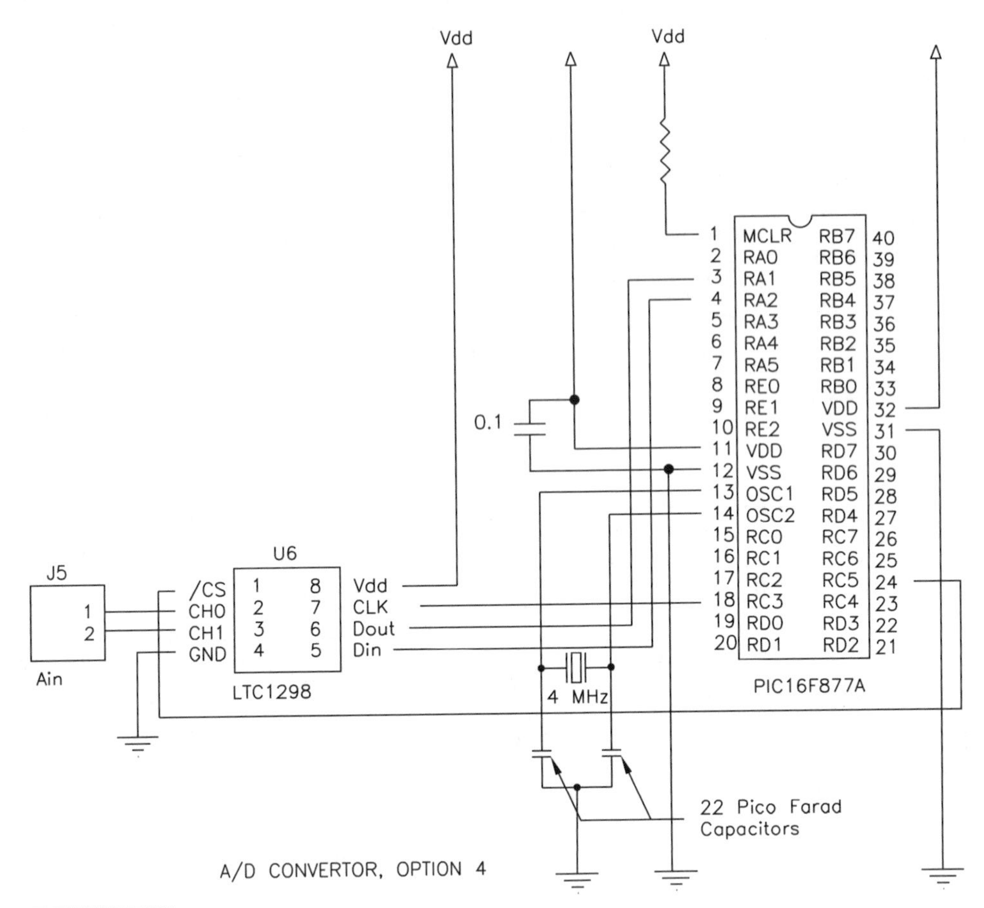

Figure 7.8 A to D converter use

very small and need to be digitized so that they can be manipulated in a digital engine. The interface that converts these small analog signals to useful digital information is the A to D converter, as shown in Figure 7.8.

microEngineering Labs provides a program on their web site that shows how to read the 12 bit LTC1298; see Program 7.4.

Program 7.4 is a PICBASIC PRO program to read LTC1298 ADC. It defines the boot loader to be used with the "Define LOADER_USED 1" instruction. Adding this instruction does not affect normal program operation.

Program 7.4 Read from 12-bit LTC1298 A to D chip.

```
DEFINE LOADER_USED 1          ;
DEFINE LCD_DREG PORTD         ; define LCD pins
DEFINE LCD_DBIT 4             ;
DEFINE LCD_RSREG PORTE        ;
```

(continued)

Program 7.4 **Read from 12-bit LTC1298 A to D chip** (*continued*).

```
DEFINE LCD_RSBIT 0                       ;
DEFINE LCD_EREG PORTE                    ;
DEFINE LCD_EBIT 1                        ;
INCLUDE "MODEDEFS.BAS"                   ; alias pins
                                         ;
CS VAR PORTC.5                           ; chip select
CK VAR PORTC.3                           ; clock
DI VAR PORTA.2                           ; data in
DO VAR PORTC.1                           ; data out
                                         ; allocate variables
ADDR VAR BYTE                            ; channel address/mode
RESULT VAR WORD                          ;
X VAR WORD                               ;
Y VAR WORD                               ;
Z VAR WORD                               ;
 HIGH CS                                 ; chip select inactive
 ADCON1 = 7                              ; set PORTA, PORTE to
                                         ; digital
 LOW PORTE.2                             ; LCD R/W line low (W)
 PAUSE 100                               ; wait for LCD to start
GOTO MAINLOOP                            ; skip subroutines
                                         ; subroutine to read a/d
                                         ; converter
GETAD:                                   ;
  CS = 0                                 ; chip select active
                                         ; send address/mode -
                                         ; Start bit, 3 bit addr,
                                         ; null bit]
  SHIFTOUT DI, CK, MSBFIRST, [1\1, ADDR\3, 0\1]
  SHIFTIN DO, CK, MSBPRE, [RESULT\12] ; get 12-bit result
  CS = 1                                 ; chip select inactive
RETURN                                   ;
                                         ; subroutine to get x
                                         ; value (channel 0)
GETX:                                    ;
  ADDR = %00000101                       ; single ended, channel
                                         ; 0, MSBF high
  GOSUB GETAD                            ;
  X = RESULT                             ;
RETURN                                   ;
                                         ; subroutine to get y
                                         ; value (channel 1)
GETY:                                    ;
 ADDR = %00000111                        ; single ended, channell,
                                         ; MSBF high
```

(*continued*)

Program 7.4 **Read from 12-bit LTC1298 A to D chip** (*continued*).

```
 GOSUB GETAD                        ;
 Y = RESULT                         ;
RETURN                              ;
                                    ; subroutine to get z
                                    ; value (differential)
GETZ:                               ;
 ADDR = %00000001                   ; differential (ch0 = +,
                                    ; ch1 = -), MSBF high
 GOSUB GETAD                        ;
 Z = RESULT                         ;
RETURN                              ;
                                    ;
MAINLOOP:                           ;
  GOSUB GETX                        ; get x value
  GOSUB GETY                        ; get y value
  GOSUB GETZ                        ; get z value
  LCDOUT $FE, 1, "X=", #X, " Y=", #Y, " Z=", #Z ; Send values
                                            ; to LCD
  PAUSE 100                         ; do it about 10 times a
                                    ; second
GOTO MAINLOOP                       ; go it forever
END                                 ; end program
```

Program 7.4 reads three values from the A to D converter and displays them as X, Y, and Z values on the LCD. The 1298 is a two-channel device, and the two signals are read from pins 2 and 3 on the device. The third value being displayed on the LCD is the differential between the two values instead of two separate signals. This means the device is being used to look at the two inputs, not as two individual inputs but as one signal across both lines.

The two channels are connected to the two pins at J5. These are the two pins that the crystal for the clocks goes across and, as mentioned before, there is a hardware conflict between using the clock chips and the A to D converter.

The LTC 1298 can provide a maximum of 11.1 thousand samples per second. The device accepts an analog reference voltage between –0.3 and Vcc +0.3 V, so the signals that are to be read have to be conditioned to reflect these requirements.

Sockets U7 and U8

Sockets U7 and U8 are designed for temperature sensing experiments.

Note *U8 is a three-hole group for soldering in a 3-wire temperature sensing device and is located next to U7.*

Program 7.5 **DS1820 (Read temperature by microEngineering Labs Inc.)**

```
  The DS1820 temperature reading device goes in socket U7, and
the DS1620 temperature sensor has to be soldered into socket
U8.Program 5 is a microEngineering Labs PICBASIC PRO program
to read the DS1820 1-wire temperature sensor and display the
temperature on the LCD.
  DEFINE LCD_DREG  PORTD               ; define lcd pins
  DEFINE LCD_DBIT  4                   ;
  DEFINE LCD_RSREG PORTE               ;
  DEFINE LCD_RSBIT 0                   ;
  DEFINE LCD_EREG  PORTE               ;
  DEFINE LCD_EBIT  1                   ;
                                       ; allocate variables
  COMMAND VAR BYTE                     ; storage for command
  I VAR BYTE                           ; storage for loop
                                       ; counter
  TEMP VAR WORD                        ; storage for temperature
  DQ VAR PORTC.0                       ; alias DS1820 data pin
  DQ_DIR VAR TRISC.0                   ; alias DS1820 data
                                       ; direction pin
                                       ;
  ADCON1 = %00000111                   ; set PortA and PortE to
                                       ; digital
  LOW PORTE.2                          ; lcd r/w line low (w)
  PAUSE 100                            ; wait for lcd to start
  LCDOUT $FE, 1, "TEMP IN DEGREES C"   ; display sign-on message
                                       ; mainloop to read the
                                       ; temp and display on lcd
  MAINLOOP:                            ;
  GOSUB INIT1820                       ; init the DS1802
  COMMAND = %11001100                  ; issue skip rom command
  GOSUB WRITE1820                      ;
  COMMAND = %01000100                  ; start temperature
                                       ; conversion
  GOSUB WRITE1820                      ;
  PAUSE 2000                           ; wait 2 seconds for
                                       ; conversion to complete
  GOSUB INIT1820                       ; do another init
  COMMAND = %11001100                  ; issue skip rom command
  GOSUB WRITE1820                      ;
  COMMAND = %10111110                  ; read the temperature
  GOSUB WRITE1820                      ;
  GOSUB READ1820                       ;
                                       ; display the decimal
                                       ; temperature
```

(continued)

Program 7.5 **DS1820 (Read temperature by microEngineering Labs Inc.)** *(continued)*

```
LCDOUT $FE, 1, DEC (TEMP >> 1), ".", DEC (TEMP.0 * 5),
                                          ; "DEGREES C"
GOTO MAINLOOP                             ; do it forever
                                          ; initialize DS1802 and
                                          ; check for presence
INIT1820:                                 ;
LOW DQ                                    ; set the data pin low
                                          ; to init
PAUSEUS 500                               ; wait > 480 µs
DQ_DIR = 1                                ; release data pin (set
                                          ; to input for high)
PAUSEUS 100                               ; wait > 60 µs
IF DQ = 1 THEN                            ;
LCDOUT $FE, 1, "DS1820 NOT PRESENT"       ;
PAUSE 500                                 ;
GOTO MAINLOOP                             ; try again
ENDIF                                     ;
PAUSEUS 400                               ; wait for end of
                                          ; presence pulse
RETURN                                    ;
                                          ; write "command" byte
                                          ; to the DS1820
WRITE1820:                                ;
FOR I = 1 TO 8                            ; 8 bits to a byte
  IF COMMAND.0 = 0 THEN                   ;
    GOSUB WRITE0                          ; write a 0 bit
  ELSE                                    ;
    GOSUB WRITE1                          ; write a 1 bit
  ENDIF                                   ;
    COMMAND = COMMAND >> 1                ; shift to next bit
  NEXT I                                  ;
RETURN                                    ;
                                          ; write a 0 bit to the
                                          ; DS1802
WRITE0:                                   ;
  LOW DQ                                  ;
  PAUSEUS 60                              ; low for > 60 µs for 0
  DQ_DIR = 1                              ; release data pin (set
                                          ; to input for high)
RETURN                                    ; write a 1 bit to the
                                          ; DS1820
WRITE1:                                   ;
  LOW DQ                                  ; low for < 15 µs for 1
  @ NOP                                   ; delay 1us at 4 MHz
```

(continued)

Program 7.5 **DS1820 (Read temperature by microEngineering Labs Inc.)** (*continued*)

```
  DQ_DIR = 1                          ; release data pin (set
                                      ; to input for high)
  PAUSEUS 60                          ; use up rest of time
                                      ; slot
RETURN                                ; read temperature from
                                      ; the DS1820
READ1820:                             ;
  FOR I = 1 TO 16                     ; 16 bits to a word
    TEMP = TEMP >> 1                  ; shift down bits
    GOSUB READBIT                     ; get the bit to the top
                                      ; of temp
  NEXT I                              ;
RETURN                                ; read a bit from the
                                      ; DS1820
READBIT:                              ;
  TEMP.15 = 1                         ; preset read bit to 1
  LOW DQ                              ; start the time slot
  @NOP                                ; delay 1us at 4mhz
  DQ_DIR = 1                          ; release data pin (set
                                      ; to input for high)
  IF DQ = 0 THEN                      ;
    TEMP.15 = 0                       ; set bit to 0
  ENDIF                               ;
  PAUSEUS 60                          ; wait out rest of time
                                      ; slot
RETURN                                ;
END                                   ; end
```

Program 7.6 is a PICBASIC PRO program written by microEngineerling Labs to read the DS1620 3-wire temperature sensor and to display the temperature on the LCD.

Program 7.6 **DS1620**

```
INCLUDE "MODEDEFS.BAS"                ; define lcd pins
DEFINE LCD_DREG  PORTD                ;
DEFINE LCD_DBIT  4                    ;
DEFINE LCD_RSREG PORTE                ;
DEFINE LCD_RSBIT 0                    ;
DEFINE LCD_EREG  PORTE                ;
DEFINE LCD_EBIT  1                    ;
                                      ; alias pins
RST VAR PORTC.0                       ; reset pin
DQ VAR PORTC.1                        ; data pin
CLK VAR PORTC.3                       ; clock pin
                                      ; allocate variables
```

(continued)

Program 7.6 **DS1620** (*continued*)

```
TEMP VAR WORD                           ; storage for temperature
LOW RST                                 ; reset the device
ADCON1 = %00000111                      ; set PortA and PortE to
                                        ; digital
LOW PORTE.2                             ; lcd r/w line low (w)
PAUSE 100                               ; wait for lcd to start
LCDOUT $FE, 1, "TEMP IN DEGREES C"      ; display sign-on message
                                        ; mainloop to read the
                                        ; temp and display on lcd
MAINLOOP:                               ;
RST = 1                                 ; enable device
SHIFTOUT DQ, CLK, LSBFIRST, [$EE]       ; start conversion
RST = 0                                 ;
PAUSE 1000                              ; wait 1 second for
                                        ; conversion to complete
RST = 1                                 ;
SHIFTOUT DQ, CLK, LSBFIRST, [$AA]       ; send read command
SHIFTIN DQ, CLK, LSBPRER, [TEMP\9]      ; read 9 bit temperature
RST = 0                                 ;
                                        ; display the decimal
                                        ; temperature
LCDOUT $FE, 1, DEC (TEMP >> 1), ".", DEC (TEMP.0 * 5),
                                        ; "DEGREES C"
GOTO MAINLOOP                           ; do it forever
END                                     ;
```

8

SERIAL COMMUNICATIONS: SOCKETS U9 AND U10

An important part of controlling the motors we are interested in controlling is getting information to and from a personal computer both for record keeping and command generations. A serial RS232 or RS485 interface provides an easy to use standard for communication between personal computers and PIC microcontrollers. In this chapter we will cover the details of how this is done.

If all you need is a quick serial communications implementation, Program 8.1 from the microEngineering Labs web site gives all the code you need to read and write to the UART (universal asynchronous receiver/transmitter). Combine the single character code in a loop to read and write more than one character (that is, strings).

Program 8.1 RS232 Communications. Communicate with a computer.

```
CLEAR                              ; read and write hardware usart
OSC 4                              ; osc speed MUST BE SPECIFIED
B1   VAR BYTE                      ; initialize usart
  TRISC = %10111111                ; set TX (PORTC.6) to out,
                                   ; rest in
  SPBRG = 25                       ; set baud rate to 2400
  RCSTA = %10010000                ; enable serial port and
                                   ; continuous receive
  TXSTA = %00100000                ; enable transmit and
                                   ; asynchronous mode
                                   ;
                                   ; echo received characters in
                                   ; infinite loop
LOOP:                              ;
GOSUB CHARIN                       ; get a character from serial
                                   ; input, if any
  IF B1 = 0 THEN LOOP              ; no character yet
  GOSUB CHAROUT                    ; send character to serial
                                   ; output
```

(continued)

Program 8.1 RS232 Communications. Communicate with a computer. (*continued*)

```
GOTO LOOP                        ; do it forever
                                 ;
CHARIN:                          ; subroutine to get a character
                                 ; from
                                 ; usart receiver
B1 = 0                           ; preset to no character
                                 ; received
  IF PIR1.5 = 1 THEN             ; if receive flag then...
    B1 = RCREG                   ; get received character to b1
  ENDIF                          ;
CIRET:                           ;
RETURN                           ; go back to caller
                                 ; subroutine to send a
                                 ; character
CHAROUT:                         ; to usart transmitter
IF PIR1.4 = 0 THEN CHAROUT       ; wait for transmit register
                                 ; empty
  TXREG = B1                     ; send character to transmit
                                 ; register
RETURN                           ; go back to caller
END                              ;
```

On the other hand, if you need a greater understanding of what is going on, the LAB-X1 allows you to experiment with two types of serial communications. The board comes with hardware for the RS232 standard on board and an empty socket that can be configured with the RS485 protocol (a line driver IC has to be added). Only one type of communications can be active at any one time, and the chip that is not being used must be removed from the board. The RS232 communications are routed to the DB-9 female connector on the board, and there are PC board holes for a 3-pin connector at J10 for RS485 communications. The IC required by the RS485 communications is the SN175176A or equivalent line driver.

The two standards are similar and simply stated: using RS485 allows you to go longer distances and the communication is more noise tolerant. This is related to the capacitance of the lines being used and stronger drivers that are employed.

The compiler supports communications to both standards and the specified compiler commands should be used whenever possible. Writing your own sequences for controlling serial communications is counterproductive, though it might be instructive. The compiler uses the same commands to access both standards, and the hardware determines how the signals are sent out and received. (Take a minute to look at the wiring diagrams in Figures 8.1 and 8.2 to see what the complications are.)

Communications are timed according to the specification of the oscillator. For the proper timing to be achieved, the OSC command has to be set to the actual oscillator frequency that is in use. If the frequencies are not matched, communications will be speeded up or slowed down (in speed) based on the extent of the mismatch. If you are actually using a 4 MHz oscillator and specify OSC 20 in your program, the communications will slow down to one-fifth the specified speed because the system is actually running at one-fifth the 20 MHz speed.

In order to experiment with communications, we need to be able to communicate with an external device. The easiest device to use is a personal computer with a dumb terminal program. The various versions of Microsoft Office Works software all contain a terminal program that you can access and use.

We can set up a dumb terminal by following these instructions:

1. From the Windows Start Menu, select Programs | Accessories | Communications | HyperTerminal.

   If for some reason HyperTerminal does not show up here, select Help from the Start menu, search for "HyperTerminal," and select Finding in 2000. This will give you a window with a link to HyperTerminal. Downloads are free.
2. Set up the HyperTerminal for:
   - 8 bits
   - No parity
   - 1 stop bit
   - 2400 baud

The wiring schematic for the connections between the PIC and the 9 pin serial connector (the DB9S) on the LAB-X1 is shown in Figure 8.1.

This is what the system defaults to with the PICBASIC PRO compiler. We will use these settings for all our experiments. Set it up and save your terminal to the desktop for easy access.

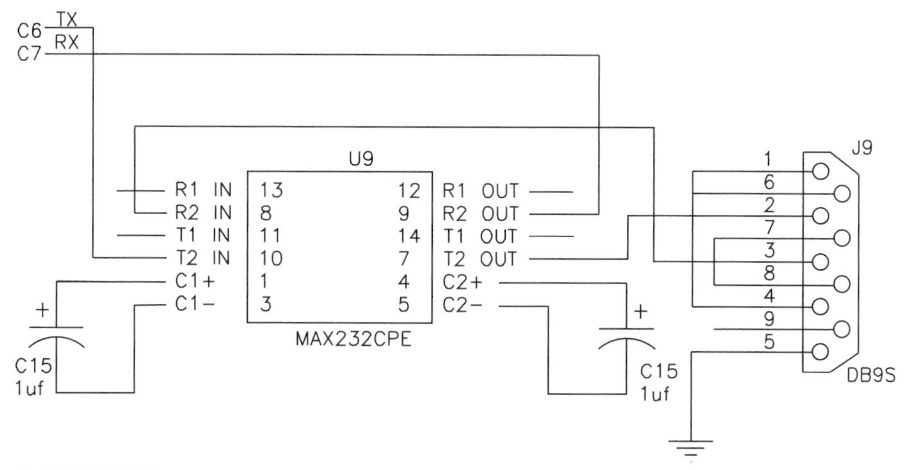

Figure 8.1 **RS232 Communications wiring: Wiring diagram for the RS232 standard**

Connect the serial cable between the computer and the MCU. Since this same cable is also used for programming the MCU, if you have only one serial port you have to disconnect it from the serial programmer after every programming session if you are using a serial programmer.

Note *There are actually two RS232 drivers on the MAX232CPE. The unconnected I/O lines belong to the second driver.*

When and How Will I Know the Interface Is Working?

Once set up properly, whatever is sent out by the LAB-X1 will show up on the HyperTerminal screen, and whatever is typed in at the computer keyboard will show up on the LAB-X1 LCD and the HyperTerminal screen.

We will be using the hardware serial output command HSEROUT, which applies to the first COM port on the LAB-X1.

Note *The LAB-X1 has only one port, so HSEROUT2 is not applicable for use with this MCU. The most obvious use for two ports is for data collection and conversion with useful filtering, where the data comes in on one port (from an instrument or what ever you are working with) and is translated, massaged, and filtered and then goes out on the other port.*

Let us write a simple program, with no safety or error correction interlocks, to send a series of 75 uppercase "A" to the computer, one "A" at a time with no delay between transmissions. Seventy-five characters will fit on one line with the carriage return. This will keep the LAB-X1 busy for about 0.25 seconds every time you press the reset pushbutton while you adjust the terminal settings, if that is necessary. The basic code needed to send the 75 characters is provided in Program 8.2. There is no code for receiving information from the computer at this juncture.

Program 8.2 RS232 Communications. Send information to the computer.

```
CLEAR                    ;
DEFINE OSC 4             ;
DEFINE HSER_RXSTA 90h    ; setting up the communications
DEFINE HSER_TXSTA 20h    ; variables
DEFINE HSER_BAUD 2400    ;
DEFINE HSER_SPBRG 25     ;
HSEROUT [$D, $A, $A]     ; a carriage return and two line feeds
ALPHA VAR BYTE           ; set counter variable
FOR ALPHA = 1 TO 75      ;
HSEROUT ["A"]            ; loop to send out the 75 'A'
                         ; characters
NEXT ALPHA               ;
END                      ;
```

For the communications protocol to work properly, we need to match the settings of the HyperTerminal. As indicated in the compiler manual, this is done with the arguments in the HSEROUT command and by the protocol-related DEFINEs in the program. These DEFINEs are shown in Program 8.2.

Next, we need to receive information from the computer and display it on the two lines of the LCD. We will set it up so that the LCD will be cleared after every 20 characters received so that we don't run out of space on line 1 of the display. The operant command for receiving the data is

```
HSERIN {ParityLabel, }{Timeout, Label,}[Item{,...}]
```

The DEFINEs in the first program segment define the variables to be used. Timeout and label are optional and are used to allow the program to continue if characters are not received in a timely manner. Timeout is specified in milliseconds. See the more detailed discussion on Timeout in the compiler manual for more information. In our case, the timeout means that the program will jump to the sending routine whenever there is nothing in the receiver buffer. The receiver buffer has preference as it's set up. However, you need to keep in mind that the receive buffer is only two bytes long, so we cannot linger on the send side too long before checking on the receive buffer again.

Things to keep in mind when receiving information:

- Certain control characters do not show up on screen.
- Some characters are not implemented.
- The receiving buffer is only two characters long.
- We have not taken any precautions for transmission/reception errors and so on, and that can get more complicated than what we need to cover at this level of our expertise.

We can write a short program to receive and display information on the LAB-X1. Since the information will be displayed on the LCD, we have to include all the usual code for accessing the LCD in our program. The code for doing all this is provided in Program 8.3.

Program 8.3 RS232 Communications. Receiving information from the computer

```
CLEAR                           ; clear the memory
DEFINE LCD_DREG   PORTD         ; define LCD registers and bits
DEFINE LCD_DBIT  4              ;
DEFINE LCD_RSREG  PORTE         ;
DEFINE LCD_RSBIT  0             ;
DEFINE LCD_EREG   PORTE         ;
DEFINE LCD_EBIT  1              ;
                                ;
CHAR VAR BYTE                   ; variables used in the routine
```

(continued)

Program 8.3 RS232 Communications. Receiving information from the computer (*continued*)

```
                                   ; storage for serial character
COL VAR BYTE                       ; keypad column
ROW VAR BYTE                       ; keypad row
KEY VAR BYTE                       ; key value
LASTKEY VAR BYTE                   ; last key storage
                                   ;
ADCON1 = %00000111                 ; set PORTA and PORTE to
                                   ; digital
LOW PORTE.2                        ; LCD R/W line low (W)
PAUSE 500                          ; wait for LCD to startup
OPTION_REG.7 = 0                   ; enable PORTB pullups
                                   ;
KEY = 0                            ; initialize variables
LASTKEY = 0                        ;
                                   ;
LCDOUT $FE, 1                      ; initialize and clear display
                                   ;
LOOP: HSERIN 1, TLABEL, [CHAR]; get a char from serial port
LCDOUT CHAR                        ; send char to display
                                   ;
TLABEL:                            ;
GOSUB GETKEY                       ; get a key press if any
IF (KEY != 0) AND (KEY != LASTKEY) THEN ;
HSEROUT [KEY]                      ; send key out serial port
ENDIF                              ;
LASTKEY = KEY                      ; save last key value
GOTO LOOP                          ; do it all over again
                                   ;
GETKEY:                            ; subroutine to get a key from
                                   ; keypad
  KEY = 0                          ; preset to no key
  FOR COL = 0 TO 3                 ; 4 columns in keypad
    PORTB = 0                      ; all output pins low
    TRISB = (DCD COL) ^ $FF        ; set one column pin to output
    ROW = PORTB >> 4               ; read row
    IF ROW != %00001111 THEN GOTKEY; If any key down, exit
  NEXT COL                         ;
  RETURN                           ; no key pressed
                                   ;
GOTKEY:                            ; change row and col to ASCII
                                   ; key number
  KEY = (COL * 4) + (NCD (ROW ^ $F)) + "0" ;
RETURN                             ; subroutine over
END                                ;
```

Next, we combine the programs to give us full communications between the LAB-X1 and the HyperTerminal program in the computer. (This is left to you; however, some pertinent hints are provided.)

The HyperTerminal software takes care of receiving, displaying, and sending characters without need for any further modification by us.

The LAB-X1 software has to receive characters from the terminal program and send them to the LCD, and it has to read the keyboard and send what it sees to the terminal. The receiving and sending has to be in the same main loop.

Using the RS485 Communications

If we want to use the more robust RS485 for communicating between a personal computer and the PIC, the wiring schematic we will need to use is provided in Figure 8.2.

In order to use the RS485 serial communications standard, pins have to be installed in JP4 to enable the ground connections and J10 to carry the communications. As was mentioned before, the RS232 IC in U9 has to be removed.

The commands and software program used will be the same as were used for the RS232 communications.

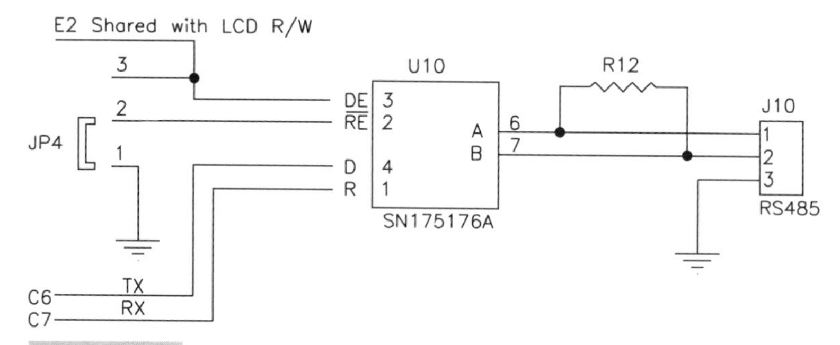

Figure 8.2 RS485 Communications wiring: Wiring diagram for the RS485 standard

9

USING LIQUID CRYSTAL DISPLAYS: AN INFORMATION RESOURCE

The use of LCD (liquid crystal display) modules is covered in great detail in this chapter because they form an important part of any project based on the use of the PIC line of microprocessors and the PICBASIC PRO compilers. We will consider the popular 2-line by 16-character display in detail, but the information is applicable to most small LCDs on the market today.

The first part of this chapter summarizes the information needed to write to the LCD, and the second part goes into much greater detail, including initializing requirements and hardware connection schemes.

The PICBASIC PRO compiler provides full support for the 2-line by 20-character display provided on the LAB-X1 board as well as for other similar displays controlled by the Hitachi HD44780U and compatible controllers. Before a display can be used, it is necessary to tell the compiler where the display is located in memory. This is done by setting the value in a number of DEFINEs that have been named and predefined in the compiler. These DEFINEs let you write to any LCD at any memory/port location in your project with the compiler. The specific DEFINEs related to the control of the LAB-X1 display are as follows:

- Identifies the port that is connected to the LCD data pins:

```
DEFINE      LCD_DREG      PORTD
```

- Decides how many bits of data to use and the starting bit (this can be a 0 or a 4 for the data starting bit and 4 or 8 for the number of bits used):

```
DEFINE      LCD_DBIT      0 (or 4)
DEFINE      LCD_BITS      4 (or 8)
```

- Specifies the register that will contain the register selection bit and the number of the bit that will be used to select the register:

```
DEFINE      LCD_RSREG    PORTE
DEFINE      LCD_RSBIT    0
```

When we transfer the data, we have to enable the transfer by toggling a bit. The port and bit for doing this are defined with the following two lines:

```
DEFINE      LCD_EREG     PORTE
DEFINE      LCD_EBIT     1
```

Decide whether we are going to read data from or write data to the LCD. This is the read/write bit and is defined with the following two lines:

```
DEFINE      LCD_RWREG    PORTE
DEFINE      LCD_RWBIT    2
```

Most of the time we do not need to read data from the LCD module, and this bit can be left low:

```
LOW PORTE.2          ; set LCD R/W low (if write only is to be
                     ; implemented)
```

If we are not *ever* going to read from the LCD module, the preceding bit can be set and left low, or it can be tied low with hardware (doing so will save one line on the PIC).

Since the PIC 16F877A has analog capability, it will come up in analog mode on startup and reset. This has to be changed to digital mode by setting the A to D control register bits. This is done with the following:

```
ADCON1 = %00000110          ; makes all of PORTA and PORTE
                            ; digital (%00000111 can also be used)
```

The LCD takes a considerable time to start up and initialize itself, so we have to wait for about 500 ms before we write to it. If there are a lot of other tasks that will take place before the first write to the LCD, this time can be reduced. (A trial and error approach can be used.)

```
PAUSE 500                   ; Wait 0.5 secs. for LCD to start up
```

Usually the first command to the LCD clears the display and writes to it on line 1 but I am showing it as two lines, where the first line clears the display and the second line positions the cursor at the first position on the first line and prints the word "Blank."

```
LCDOUT $FE, 1                    ; clear the LCD
LCDOUT $FE, $80, "Blank"         ; written to line 1 position 1
                                 ; of the LCD
```

All commands (as opposed to characters) sent to the LCD are preceded by the code $FE or decimal 254. The basic commands needed to write to the LCD are listed in Table 9.1.

All these and other codes are described in detail in the Hitachi data sheet. You must learn where in the data sheet these listings are located so you can refer to them when necessary.

The commands in Table 9.1 apply to all LCDs using the Hitachi HD44780U controller or an equivalent. See the data sheet for this controller for more detailed information. This controller has a lot of commands not shown here, including limited graphic capability within the characters.

It is useful to have the full 40-plus page data sheet on hand when you are doing anything more than sending text to the LCD. The Hitachi 44780 data sheet can be downloaded at no charge from the following URLs:

```
http://semiconductor.hitachi.com/products/pdf/99rtd006d2.pdf
http://pic.rocklizard.org/LCDDriver/HD44780U.pdf
```

LCD displays require that each line be addressed with its own starting position as indicated previously. The exception is that most 16-character single line displays are designed such that the first eight characters start at $80 and the next eight start at $C0. The 16 characters appear to be on two lines to the controller, though they are displayed as one line on the LCD module. Please also note that lines 3 and 4 of 4-line displays also have an irregular addressing scheme.

If more characters than can be displayed on a line are sent to the LCD, they will be stored in memory space in the LCD. They can be scrolled back across the screen if needed. The number of characters that an LCD can store in its display memory are a property of the LCD as determined by the manufacturer. You can also scroll the display up and down if you design the commands and write the software to do so. This is not built into the LCD or the controller software.

TABLE 9.1 LCD CODE LISTINGS

$FE,	**$01**	Clear display. (An uncleared display shows dark rectangles in all the spaces.)
$FE,	**$02**	Go to home. Position 1 on line 1.
$FE,	**$0C**	All cursors off. This is the default condition on startup.
$FE,	**$0E**	Underline cursor on.
$FE,	**$0F**	Underline cursor off. This is the default condition on startup.
$FE,	**$10**	Mover cursor right one position.
$FE,	**$14**	Mover cursor left one position.
$FE,	**$80**	Move cursor to position 1 of line 1.
$FE,	**$C0**	Move cursor to position 1 of line 2.
$FE,	**$94**	Move cursor to position 1 of line 3.
$FE,	**$D4**	Move cursor to position 1 of line 4.

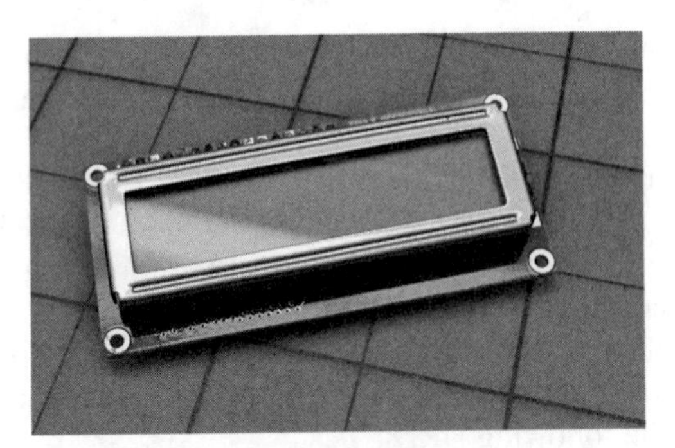

Figure 9.1 2-line by 16-character LCD module: a 1- or 2-line LCD can add a tremendous amount of utility to a project.

Figure 9.1 illustrates a typical 16-character by 2-line LCD display. These units can be purchased for approximately $6.00 (circa 2009) each on the Internet.

Using LCDs in Your Projects

It is generally agreed that most projects benefit from having a 1- or 2-line display incorporated into them. However, these displays tend to be rather pricey (about $50) when provided with the necessary controlling IC and quite reasonable (about $5 to $10) when bought without the supporting package. Since a PIC microcontroller can be purchased for about $5, we should be able to have a complete display unit for a marginal cost of about $10 if we can figure out how to program your PIC microcontroller to control the display.

The readily available and inexpensive 2-line by 16-character LCD ($6 at All Electronics and often less on the Internet) offers us the ability to display information in a limited but useful way in your projects. Mastering the use of this LCD display means that we have gained the expertise to write any character (Hitachi standard or designed by you) at any location, at any time, or in response to any event whenever we want. This chapter addresses this problem in detail and tells you what you need to know and do to make these inexpensive displays a part of all of your projects.

In this chapter you will learn how to control an LCD. The code that you create can be incorporated into almost any PICBASIC PRO program and will control the LCD from any available half port (nibble) and three other free I/O lines. The code will be more linear than it needs to be so that you can see exactly what is going on. Once you understand what needs to be done, you can write more compact and sophisticated code that will get the job done in the way that you want.

Understanding the Hardware and Software Interaction

The hardware that we are considering consists of an LCD with an integral controller that is incorporated into the display by the display manufacturer. In this particular case, this is the Hitachi HD44780U controller. Displays are available without this or any other controller, but controlling displays without a controller is way beyond the scope of this workbook. For our projects, be sure you buy only those units that have this controller built in as a part of the display.

Controlling the display consists of telling this controller what we want it to do. The instructions are easy to understand and allow you to control each and every pixel and all the functions that the display can perform with relative ease. You do not have to read or understand the rather extensive 40-plus page data sheet that Hitachi provides for this controller, but it is well worth the trouble to download the data sheet and study it. You do not *need* to download this file either but you should know where to find it if you need it. We will go over almost everything that you need to know to control the display as a part of this exercise. Having said this, I strongly encourage you to get familiar with what this controller can do in great detail.

In the LCD display the imbedded controller provides the interface between the user and the display. The controller of choice for almost all the LCDs on the market is the Hitachi HD44780U. This very powerful controller gives you complete control over the LCD. It allows you to address each and every pixel on the display, It also has a built-in set of ASCII characters for use by the user. Our task is to learn how to use this controller to put what we want, when we want, where we want, in the display.

Note *The other common controller is the Epson SED series controller. Its operation and instruction set is very similar to that of the Hitachi controller. We will not consider the EPSON or any other controller in this book.*

You will find that almost all the smaller liquid crystal displays on the market are controlled by the Hitachi controller. This means that once you learn to control one display, you will be able to control most of them with the code that you create. As a matter of fact, we will write the code in a way that will be universal in its application. We will define variables such as the number of characters spaces and number of lines in the display as a part of the program setup.

Note *It is also useful and usually easier to use the DEFINES that are created by the PBP compiler to control the display*

The addresses of the local memory locations (the ones in the LCD) used by the LCD have already been fixed, as has the instruction set that we use to write to the LCD, so we do not have to create any of this rather sophisticated code.

As stated previously, you do not need the full data sheet, but you do need to know the basic command set that controls the data transfer to your particular display. This is usually provided by the organization that you buy the LCD from, and consists of two or three pages. You will need to refer to the data sheet only if you want to create special characters or bar graphs and the like on the display. The control that the Hitachi controller provides is very comprehensive, but you don't need to be familiar with it to use a display effectively. Most of what you need to know for day to day use will be covered in the next exercise. See Table 9.1 for the basic commands that are needed to provide day to day control of a display.

Talking to the LCD

The control codes in Table 9.1 allow you to configure the display, set display parameters, set the shape and position of the cursor and so on. To differentiate them from the character commands, each control code must be preceded by a hexadecimal FE or a decimal 254 to tell the controller that the next character sent to the display will be a control code. After receiving one control code and its argument, the Hitachi controller resets to the data mode automatically.

The controller supports the ASCII standard. All uppercase and lowercase characters and numerals are supported, as are punctuation marks and the standard text support characters.

It is also possible to design your own font for use with the displays, though 5 by 7 (or even 10) dots and two lines do limit what can be done. All the information you need to do so is in the Hitachi HD44780U data sheets. Greek characters and certain scientific notations are useful for most scientific applications.

THE HARDWARE CONNECTIONS

Let us take a closer look at the LCD to determine how you might wire it to the MCU.

Study the data sheet that came with the LCD. Find the pinout descriptions and study them. The 16 pins are usually identified as shown in Table 9.2.

Communications between the LCD and the PIC can use either all eight pins of port D or only the pins from D4 to D7. This is explained in detail in the PICBASIC PRO language manual.

Looking at the information provided with the 16-character by 2-line displays, we find that the control implementation can take place if we have a port and a few lines available to control the LCD. It does not have to be controlled from any predefined lines; we can select all the lines that are needed to support the display in the project, and they can be on any available port. The only requirement seems to be that the four/eight data line be either the contiguous top or the bottom half of a port. This is not a particularly demanding requirement; it means that the smaller PICs cannot be used if we will need a lot of I/O lines in our project. The other three lines that are needed can be on any of the other ports and do not all need to be on the same port. Since we are considering only one PIC in this workbook, this will mean that we will use the 16F877A. I have included the circuitry and code for this in the next exercise so that you can see exactly what needs to be done.

For now, to keep it simple we'll use the following:

Use lines 1 to 3 of PORTA as the control lines
Use PORTB as the data lines.

We are using these two ports in this way because the smaller more inexpensive PICs have only PORTA and PORTB on them. (PORTE that is used in other programs in this book is provided only on the larger PICs like the 16F877A.)

TABLE 9.2 LCD CODE LISTING: PINOUT IDENTIFICATION OF THE LCD PINS FROM THE DATA SHEET

PIN NO.	SYMBOL	DESCRIPTION	NOTES
1	VSS	Logic ground	
2	VDD	Logic power 5V	
3	VO	Contrast of the display, can usually be grounded	
4	RS	Register select	These are the 3 control lines
5	R/W	Read/write	
6	E	Enable	
7	DB0		
8	DB1		
9	DB2		
10	DB3		
11	DB4		
12	DB5		
13	DB6		
14	DB7		
15	BL	Backlight power	These two lines can be ignored
16	BL	Backlight ground	

We can redefine these to be more rational addresses whenever we want and none of the programming will have to change. Just define what ports and lines you want to use at the top of the program, and the aliases assigned to them will identify them as needed.

With this in mind, let us create the software to control a 2-line by 16-character display. Once we are happy with what we have created, we can migrate the code to other microcontrollers.

Note *You could use a 1-line (about $5) display, but that would inhibit learning how to go to line 2 and scrolling the display up and down. To experiment with these features you need to have a display with at least two lines.*

SETTING OUT THE DESIGN INTENT

We need to have the following goals in mind as we go about designing the system for controlling the LCD display.

- Control the 16 × 2 display with a PIC 16F877A microcontroller.
- Design the software so it can be an integral part of the software for any project.

- Use standard control codes so that the project is a virtual plug-in replacement for other displays and in other PICs (only minor modifications, if any, will be required).
- Use a minimum number of external components so this is a software project that can move between PIC controllers of all descriptions. All we have to do is to include the code in your project and connect the display to the selected ports.
- Use the project's regulated 5-volt power supply for everything.

Note *The PIC 16F877A has 33 I/O lines. The display will use 7 of them, so we will have 26 lines left over for the project. Since we don't need all these lines, we could have used the 16F84A. When we move to the PIC 16F84A, no program changes should be needed, other than changing the line and port addresses in the defines.*

Materials Needed

We will need the following hardware for this project:

- An experimental solderless breadboard
- A PIC microcontroller, 16F877A or 16F84A
- One bare 2-line by 16-character display module with a Hitachi controller
- One 4 MHz crystal
- Two 22 pf capacitors
- A regulated 5-volt power supply from the breadboard
- One 470 ohm 1/4-watt resistor
- Some 22 gauge insulated single-strand hookup wire
- 1K ohm 1/8-watt pull up resistor
- In addition, you will need the following programmer, software, and information: The microEngineering serial programmer, parallel programmer, or the USB programmer
- The PICBASIC PRO compiler and book
- The LCD data sheets that came with the LCD
- The PIC 16F877A data sheet (or the 16F84 data sheet)

We will go through the software a step at a time. After we finish, it will be your job to clean up the software and speed up its operation—that is, you will need to optimize it for the microcontroller that you are going to use in your projects.

We have to pick a specific display to work with so that we can develop real, working software for it. The display I picked sells for $6 or less and is available with a data sheet from All Electronics. AZ Displays also sells one that seems to be identical, Model ACM 1602K. The short form data sheets are similar, but the AZ one is in crisp PDF format and can be downloaded from their web site for free. Doing so will mean that you can have this information open in a window on your computer.

First we need to investigate how many pins we will need on our PIC microcontroller to interact with our display. This is summarized in Table 9.3.

Studying Table 9.3 indicates that the microcontroller does not need to be connected to lines 1, 2, 3, 15, and 16 because these have to do with power connections

TABLE 9.3 LCD PINOUTS: PINOUT IDENTIFICATION OF THE LCD PINS FROM THE DATA SHEET

PIN NO.	SYMBOL	LEVEL	DESCRIPTION AND NOTES
1	VSS	0V	Ground for logic supply
2	VDD	5.0V	Logic power supply, regulated
3	VO	—	LCD contrast, can be grounded
4	RS	H/L	H: data code; L: instruction code
5	R/W	H/L	H: read mode; L: write mode; can be tied low in hardware
6	E	H, H>L	Enable signal, pulsed from H to L, hold at H.
7	DB0	H/L	Data bit 0
8	DB1	H/L	Data bit 1
9	DB2	H/L	Data bit 2
10	DB3	H/L	Data bit 3
11	DB4	H/L	Data bit 4
12	DB5	H/L	Data bit 5
13	DB6	H/L	Data bit 6
14	DB7	H/L	Data bit 7
15	BL	—	Plus 5V; power for back lighting the display
16	BL	—	Ground for back lighting power

and not data I/O. In this case, we will be using an 8-bit data buss, and the connection to the PIC 16F877A will be as shown in Table 9.4.

We can get by with 11 lines. It might seem that it can be back to 10 if we decide to do without the ability to read from the display memory, but this is not usually so because there are times when we need to set this line high for reading the LCD's busy flag to minimize the time used by LCD routines. However, we can add about a 20 ms delay to take care of the busy time if you are really short on lines. We also need to be able to read the display memory under certain circumstances. Since this is true for all applications, we have to stay with the 11 lines for 8-bit control. (We would not have to read the display if we kept track of what we had put in the display somewhere else in the program.)

The eight data lines form a convenient byte, and we can assign one of the ports not being used for anything else. This leaves three lines:

- The E line, which needs to be toggled to transfer data to the LCD
- The RS line, which selects the register
- The (R/W), which sets the read/write status of the operations

TABLE 9.4 LCD PINOUTS: LCD PINS CONTROLLED BY THE MICROPROCESSOR

PIN NO.	SYMBOL	PIC PIN	DESIGNATION
~~1~~	~~VSS~~	-	~~Not connected to the PIC~~
~~2~~	~~VDD~~	-	~~Not connected to the PIC~~
~~3~~	~~VO~~	-	~~Not connected to the PIC~~
4	RS	1	RA2 PortA
5	R/W	2	RA3 "
6	E	18	RA1 "
7	DB0	6	RB0 PortB
8	DB1	7	RB1 "
9	DB2	8	RB2 "
10	DB3	9	RB3 "
11	DB4	10	RB4 ")
12	DB5	11	RB5 ") Half the port can also be used
13	DB6	12	RB6 ") See PICBASIC PRO manual
14	DB7	13	RB7 ")
~~15~~	~~BL~~	-	~~Not connected to the PIC~~
~~16~~	~~BL~~	-	~~Not connected to the PIC~~

Using eight lines for data allows us to generate all the codes and all the characters that the chip has in its memory. More importantly, it allows the data transfer in one step. We can also use a 4-bit protocol and transfer half a byte at a time. Using four lines for the control scheme means that the LCD can be controlled from just one port (seven lines will be used and we will still have one line to spare). The data sheets say that whether we use four lines or eight, they all have to be part of one port, and if we are using four lines they have to be the contiguous four high or the contiguous four low bits of a port—that is, we cannot use any random lines for the data bus. If we want to design our own protocol, the data transfer for the 4-bit protocol has to use the four high bits on the LCD, and we have to send the data from the PIC to the display with the high data nibble first and the low data nibble last. This little gem is not spelled out in the instructions, but this is what has to be done.

The data sheet also says that the display initializes itself on power up. We can reinitialize it under our control, but it is done automatically on startup and we cannot inhibit it. Just do nothing for about half a second and the self initialization will complete. The busy flag is set high during startup and initialization, but is indeterminate immediately after initialization starts and for 16.4 ms after the supply voltage reaches 4.5 volts. This means we cannot determine how long we'll have to wait after powering up to start doing what we want. We will set a 0.5 s wait/pause in our programs at startup; if that is not long enough, we will come back and increase the waiting time. Wait time is a must. If you do not wait, the system will not start up properly.

Automatic initialization sets the following conditions for the display:

- Display cleared
- Set for a 8-bit interface
- Set for 1 line of display
- Set for a 5 × 7 dot matrix display
- Display is turned off
- Cursor is turned off
- Blink is turned off
- Increment between characters is set to 1 (cursor moves over 1 space automatically)
- Shift is off

The preceding list may not be exactly what we want, so we will go through an initialization sequence as specified by the instructions. We do not have to go through all the steps, but we will so that we have a complete record of what needs to be done for future projects.

The instructions say that the six instructions listed in Table 9.5 have to be sent to the display during an initialization sequence. The first three instructions are identical but require different waits after each one is sent to the display.

These instructions are commands rather than data, so the RS (Register Select) line has to be held high while we initialize.

TABLE 9.5 LCD STARTUP: SEQUENCE STEPS AND TIMING DELAYS

STEP	RS LINE	R/W LINE	DATA BYTE	ENABLE LINE
1	1 high	0 low	0011xxxx	Toggle H to L
1A	Wait for 4.1 ms			
2	1 high	0 low	0011xxxx	Toggle H to L
2A	Wait for 100 μs			
3	1 high	0 low	0011xxxx	Toggle H to L
3A	Wait for 1 ms			
4	1 high	0 low	001110xx	Toggle H to L
5	1 high	0 low	00000001	Toggle H to L
6	1 high	0 low	00000110	Toggle H to L

The six lines of code listed in Table 9.5 are explained here in detail:

```
"x" in a bit = don't care
00110000   ; code to initialize the LCD (this is entered
           ; 3 times, 1st time)
           ; load for a command function
           ; wait at least 4.1 ms
           ;
00110000   ; code to initialize the LCD 2nd time
           ; load for a command function
           ; wait at least 100 µs
           ;
00110000   ; code to initialize the LCD 3rd time
           ; load for a command function
           ; wait at least 1 ms
           ;
00111000   ; put in 8 bit mode, 2 line, 5X7 dots
           ; 0
           ; 0
           ; 1 = req'd
           ; 1 = 8 bit data transfer
           ; 1 = 2 lines of display
           ; 0 = 5x7 display
           ; x
           ; x
           ; load for a command function
           ;
00010100   ; set cursor shift etc
           ; 0
           ; 0
           ; 0
           ; 1 = req'd
           ; 0 = cursor shift off
           ; 1 = shift to right, or left (0)
           ; x
           ; x
           ; load for a command function
           ;
00001111   ; LCD display status, cursor, blink etc
           ; 0
           ; 0
           ; 0
           ; 0
           ; 1 = req'd
           ; 1 = display on
           ; 1 = cursor on so we can see it
           ; 1 = blink on so we can see it
           ; load for a command function
           ;
```

```
00000110    ; lcd entry mode set, increment, shift etc
            ; 0
            ; 0
            ; 0
            ; 0
            ; 0
            ; 1 = req'd
            ; 1 = increment cursor in positive dir
            ; 0 = display not shifted
            ; load for a command function
```

At the end of these instructions the display will have been initialized to the way we want it.

There is also this business about the busy flag. The display takes time to do whatever we ask it to do, and the time varies with each task. We can wait a few milliseconds between instructions to make sure it has had enough time for the task to complete, or we can monitor the busy flag and, as soon as it is not busy, we can send the next instruction. Since time is always at a premium and we want to run as fast as we can, it means we have to consider monitoring the busy flag.

The Busy Flag

The instruction sheet says that the busy flag is bit 5 at location 11100011 in the LCD. Metacode for waiting for the busy bit in the LCD to clear is as follows:

```
Busycheck:
    Read busybyte
    Isolate busybit
    If it is busy then goto Busycheck
Return
```

You isolate the bit and if it is not low, you read the flag byte again. You do this again and again till the bit goes low. As soon as it does, you can write to the LCD and go on with the program.

We also want our display to be compatible with code generated by the PICBASIC PRO compiler. The instructions for the LCDOUT command say that the compiler would prefer that the hardware was set up for the following conditions:

- Four data bits DB4 to DB7 connected to PORTA.0 to PORTA.3
- Chip enable at PORTB.3
- Register select at PORTA.4
- Two lines of display are assumed

If we cannot meet these requirements, we have to set the addresses out as DEFINEs in each and every program that we write (or we can use an INCLUDE statement that includes a program that does this for us). It's only a few lines of code, but we will have to add the code every time, and it compromises compatibility with other systems that will no doubt be set up to meet the compiler standard.

It may turn out that the microcontroller you choose will need to have this done in any case.

The next thing we need to decide is where we are going to use the software: as an integral part of a program that is running on a larger microcontroller, where we can use all ten address lines for the display, or on a smaller dedicated microcontroller that will need only one serial line to control the display but will have to be added to the total project as a part of hardware that we design? For now, let us agree that we will go with a dedicated controller just to run the display. The software for running on a larger controller will be a subset of what we develop, so no work is lost here.

The task on the input side is to design the software that will take the serial information received on one pin and output it as 4-bit characters to the LCD with the select, read/write, and enable lines. The work needed to read the data in is done by the compiler with the SERIN instruction.

A program that does just this is provided by microEngineering Labs on their web site. Here it is as Program 9.1.

Program 9.1 **For a PIC 16F84A simulate back pack (by microEngineering Labs)**

```
DEFINE LCD_DREG   PORTD ;
DEFINE LCD_DBIT   4     ;
DEFINE LCD_RSREG PORTE  ;
DEFINE LCD_RSBIT  0     ;
DEFINE LCD_EREG   PORTE ;
DEFINE LCD_EBIT   1     ;
                        ;
CHAR VAR BYTE           ; storage for serial character
MODE VAR BYTE           ; storage for serial mode
RCV  VAR PORTB.7        ; serial receive pin
BAUD VAR PORTA.0        ; baud rate pin - 0 = 2400, 1 = 9600
STATE VAR PORTA.1       ; Inverted or true serial data
                        ; - 1 = true
                        ;
ADCON1 = %00000111      ; set PORTA and PORTE to digital
LOW PORTE.2             ; LCD R/W line low (W)
PAUSE 500               ; wait for LCD to startup
                        ;
MODE = 0                ; set mode
IF (BAUD == 1) THEN     ;
  MODE = 2              ; set baud rate
ENDIF                   ;
                        ;
IF (STATE == 0) THEN    ;
  MODE = MODE + 4       ; set inverted or true
ENDIF                   ;
                        ;
```

(continued)

Program 9.1 **For a PIC 16F84A simulate back pack (by microEngineering Labs)** (*continued*)

```
LCDOUT $FE, 1               ; initialize and clear display
                            ;
LOOP:                       ;
  SERIN RCV, MODE, CHAR     ; get a char from serial input
  LCDOUT CHAR               ; send char to display
GOTO LOOP                   ; do it all over again
END                         ; end
```

This program is for the 16F84A, but it can be used on the 16F877A with appropriate DEFINEs. You have set these DEFINEs many times before, so it should not be a problem.

If you load Program 9.1 into a PIC 16F84A, you can connect the 16F84A to the LCD. Any serial information that comes in on PORTB.7 will be displayed on the LCD. Now you can control the LCD from one line on the main processor. (The selected pin does not have to be PORTB.7; any free pin can be specified as the input data pin in the program.)

See Figure 9.2 for the wiring diagram for the 16F84A.

Note *The data does not have to be on pin B7; it can be programmed to come in at any free line. You set the line you want to use in Program 9.1.*

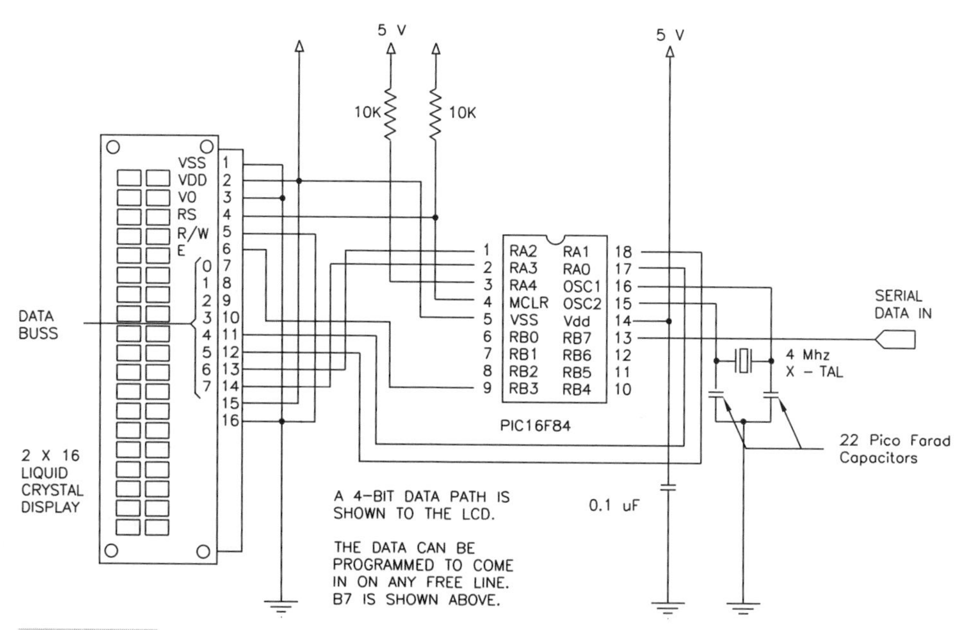

Figure 9.2 **Wiring diagram: LCD backpack using a PIC 16F84 with three lines (PWR, GND, signal).**

Liquid Crystal Display Exercises

These exercises are to be performed on the LAB-X1 board

1. Write a program to put the 26 letters of the alphabet and the 10 numerals in the 40 spaces that are available on line 1. Put four spaces between the numbers and the alphabet to fill in the four remaining spaces. Once all the characters have been entered, scroll the 40 characters back and forth endlessly though the 20 spaces that are visible on line 1.
2. Write a program to bubble the 26 uppercase letters of the alphabet through the numbers 0 to 9 on line 2 of the LCD. (In other words, first put the numbers on line 2. Then "A" takes the place of the "0" and all the numbers move over. Then the "A" takes the place of the "1" and the "0" moves back to position 1, and so on till it gets past the 9. Then the "B" starts its way across the numbers and so on.)
3. Write a program to write the numbers 0 to 9 upside down on line 1. Wait for a second and then flip the numbers right side up one by one. Provide a time delay between changes. Loop.
4. Write a program to identify the button pressed on the button pad by displaying its row number on line 1 and its column on line 2. Identify each line so you know what is being displayed where. Scroll the two lines up every time a button is pressed. Add delays in the scroll so you can actually see the scrolling take place.

Table 9.6 is for the Hitachi controller. Almost all LCDs use this scheme. The table summarizes the instructions that have to be sent to the LCD to initialize it and set its display properties along with the timing requirements for each instruction.

TABLE 9.6 LCD CODE TABLE

COMMAND	R	R	- - - - -	-	-	DATA BUSS	-	-	- - - - -		DESCRIPTION	EXECUTING
	S	W	7	6	5	4	3	2	1	0	Fosc=250 KHz	TIME
Clear Display	0	0	0	0	0	0	0	0	0	1	Clears Display and Returns to Address 0.	1.64 ms
Cursor at Home	0	0	0	0	0	0	0	0	1	x	Returns Cursor to Address 0. Also returns the display being shifted to the original position. DDRAM contents remain unchanged.	1.64 ms

(continued)

TABLE 9.6 LCD CODE TABLE (*CONTINUED*)

COMMAND	RS	RW	7	6	5	4	3	2	1	0	DESCRIPTION Fosc=250 KHz	EXECUTING TIME
Entry Mode Set	0	0	0	0	0	0	0	1	1/D	S	I/D: Set Cursor Moving Direction I/D=1: Increment I/D=0: Decrement S: Specify Shift of Display S=1: The display is shifted S=0: The display is not shifted	40 μs
Display on/off Control	0	0	0	0	0	0	1	D	C	B	Display D=1: Display on D=0: Display off Cursor C=1: Cursor on C=0: Cursor off Blink B=1: Blink on B=0: Blink off	40 μs
Cursor / Display Shift	0	0	0	0	0	1	S/C	R/L	x	x	Moves cursor or shifts the display w/o changing DD RAM contents S/C=0: Cursor Shift (RAM unchanged) S/C=1: Display Shift (RAM unchanged) R/L=1: Shift to the Right R/L=0: Shift to the Left	40 μs
Function Set	0	0	0	0	1	DL	N	F	x	x	Sets data bus length (DL), # of display lines (N), and character fonts (F.) DL=1: 8 bits F=0: 5x7 dots DL=0: 4 bits F=1: 5x10 dots N=0: 1 line display N=1: 2 lines display	40 μs
Set CG RAM Address	0	0	0	1	Character Generator (CG) RAM Address						Sets CG RAM address. CG RAM data is sent and received after this instruction	40 μs
Set DD RAM Address	0	0	1	Display Data (DD) RAM Address / Cursor Address							Sets DD RAM address. DD Ram data is sent and received after this instruction.	40 μs
Busy Flag / Address Read	0	1	BF	Address counter used for both DD & CG RAM							Reads Busy Flag (BF) and address counter contents	40 μs
Write Data	1	0	Write Data								Writes data into DDRAM or CGRAM.	46 μs
Read Data	1	1	Read Data								Reads data from DDRAM or CGRAM.	46 μs

The analog and digital properties of the various pins on PORTA and PORTE are controlled by the contents of the ADCON1 register. How this register affects these values is summarized in Table 9.7.

Note the following details about Table 9.7:

- /x/ is used to indicate the missing A.4 line.
- Pin PORTA.4 is not included because it is an open collector.
- We should pay special attention to how the reference voltages are specified at the various pins. See the data sheet.
- Settings 0110 and 0111 have identical results but are both shown so all 16 combinations will be seen. The data sheet shows 011X for both lines.
- Be aware that AN7 to AN0 are the seven analog inputs identifications, and Port E.2 to E.0 and A.5 to A.0 are pin identifications on the PIC. Do *not* confuse the *pins* with the analog *inputs*.

The settings in Table 9.7 are described in detail on page 112 of the data sheet; Table 9.7 is a short form of that table. We do not have I/O access to lines A.7 and A.6

TABLE 9.7 DIGITAL/ANALOG SELECTIONS MADE WITH THE ADCON1 REGISTER

Port/pin	E.2	E.1	E.0	A.5	/x/	A.3	A.2	A.1	A.0			
ANALOG	AN7	AN6	AN5	AN4		AN3	AN2	AN1	AN0	VREF+	VREF	CH/REF
0000	A	A	A	A	/x/	A	A	A	A	Vdd	Vss	8/0
0001	A	A	A	A	/x/	Vref	A	A	A	RA3	Vss	7/1
0010	D	D	D	A	/x/	A	A	A	A	Vdd	Vss	5/0
0011	D	D	D	A	/x/	Vref	A	A	A	RA3	Vss	4/1
0100	D	D	D	A	/x/	A	D	A	A	Vdd	Vss	3/0
0101	D	D	D	D	/x/	Vref	D	A	A	RA3	Vss	2/1
0110	D	D	D	D	/x/	A	D	D	D	Vdd	Vss	0/0
0111	D	D	D	D	/x/	A	D	D	D	Vdd	Vss	0/0
1000	A	A	A	A	/x/	Vref	Vref	A	A	RA3	RA2	6/2
1001	D	D	A	A	/x/	A	A	A	A	Vdd	Vss	6/0
1010	D	D	A	A	/x/	Vref	A	A	A	RA3	Vss	5/1
1011	D	D	A	A	/x/	Vref	Vref	A	A	RA3	RA2	4/2
1100	D	D	D	A	/x/	Vref	Vref	A	A	RA3	RA2	3/2
1101	D	D	D	D	/x/	Vref	Vref	A	A	RA3	RA2	2/2
1110	D	D	D	D	/x/	D	D	D	A	Vdd	Vss	1/0
1111	D	D	D	D	/x/	Vref	Vref	D	A	RA3	RA2	1/2

on the PIC 16F877A because they are internal to the processor. They can, however, be read and are related to the use of the parallel port capability of the PIC.

Since we have the LCD connected to PORTE, we need its pins to be digital. At this time the status of the other lines is not of interest. We can make PORTE's pins digital by selecting line 4 or 6 in Table 9.7. For our purposes, which include using the LCD, all other lines can be digital. A useful selection is line 4, which makes all of PORTA analog and all of PORTE digital. PORTA.3 can be used as the reference voltage to which the incoming signals can be compared. We would set ADCON1 as follows:

```
ADCON1=%00000011 or ADCON1 =3
```

Note *ADCON1=%00000111 is used in the programs all over this text to match what microEngineering Labs uses in their programs. It sets all the A and E lines to digital.*

Notice that the way we wire in any sensor is often identical to the way we wire in a potentiometer. In either case, the device is placed between the high and low power supply rails and we read what is the equivalent of the wiper. This is the standard way of reading a voltage into a PIC microcontroller. If other than 0 to 5 volts are to be read in, appropriate voltage dividers and the necessary safety precautions have to be provided. In this case, we can use Pin A.3 as the reference voltage, and the voltage on this pin can be adjusted with a potentiometer provided for this purpose. The pins that the reference voltages are impressed on have to be selected as indicated in Table 9.7 (only A.2 and A.3 can be used).

Part II

RUNNING THE MOTORS

10

THE PIC 18F4331 MICROCONTROLLER: A MINIMAL INTRODUCTION

There are four versions of this PIC family: 18F2331, 18F2431, 18F4331, and 18F4431. We will be using the 18F4331 because it is one of the two MCUs that will fit in the 40-pin ZIF socket of the LAB-X 1 and we do not need 16K of memory (as is provided with the PIC 18F4431). See the data sheet for the other details of the differences between the devices. The salient comparative features of the four MCUs are listed in Table 10.1

Note *The PIC 18F4331 will be used only for running the motors with encoders. All other motors will be run with the more general purpose 16F877A, which was covered in detail in Part I of this book. In general, the 16F877A and the 18F4331 are similar in many details, but you must check each feature that you want to use on the data sheet to make sure. Subtle differences can cause headaches if you are careless.*

TABLE 10.1 PROPERTY DIFFERENCES BETWEEN THE MCUs

MCU	PINS	MEMORY	DATA	EEPROM
PIC 18F2331	28 pin DIP	8K	768	256
PIC 18F2431	28 pin DIP	16K	768	256
PIC 18F4331	40 pin DIP	8K	768	256
PIC 18F4431	40 pin DIP	16K	768	256

This introduction to the MCU is necessarily minimal because we are only interested in is the motor running capability of this MCU as related to reading the encoder and generating PWM control signals. These are the aspects that will be discussed. Other features are very similar to the 16F877A. To really understand the full power of this MCU, you need to understand the 400 or so page data sheet, which is not a minor undertaking and is beyond the scope of this book.

The PIC 18F4331 Can Be Used in the LAB-X1

The PIC 18F4331 is a 40-pin integrated circuit that will fit in the ZIF socket provided on the LAB-X1 and be compatible with the power, oscillator, and other signals as needed to be made operational on the board.

The main feature of the PIC 18F4331 that we are interested in is its ability to keep track of the signals from a standard quadrature-generating encoder completely automatically without any intervention from the user. Since it is a little more than tricky to watch the encoder and run a sophisticated program at the same time with a microcontroller, this is a major benefit. Added to this, the PIC 18F4331 can be run at 40 MHz, which is twice as fast as the 16F877A and many other of the earlier MCUs in the microchip line-up. However, this cannot be done in the LAB-X1, which is limited to 20 MHz. We have been using the LAB-X1 card at 4 MHz, so we have to change the jumpers at the oscillator to change it to 20 MHz for the encoded servo motor experiments.

You can plug the PIC 18F4331 into the LAB-X1 and, if you set the MCU up to do so, it will run almost all the programs that we have discussed so far with minimal modifications. Almost all of the features of the 16F877A are available in the PIC 18F4331. However, there are differences in the setups required for some of the programs. We will not go into the details and will concentrate on running the motors instead.

There are a number of things that need to be brought to your attention right off the top when we start to talk about running motors.

Running as Fast as Possible This PIC can run at 40 MHz. If possible, it should be run at this speed. However, the LAB-X1 board can be run at up to 20 MHz only, so we have to live with using the 18F4331 at 20 MHz. Change the jumpers on the LAB X-1 to run at its maximum speed; The jumpers need to be moved to 2-1, 2-1, 3-2.

PORTA Notes In the way we are using the PIC 18F4331, lines A.3 and A.4 are connected to the encoder input signals. As such, these lines cannot be used for anything else. In the LAB-X1 line, A.3 is connected to the third potentiometer, and this can interfere with the reading of the encoder. If you set this potentiometer to its middle position, the signal from the encoder will be able to pull this line up and down without difficulty. If this is not done, it will show up as the PIC not reading the encoder. (This also means that we cannot use the A.3 line for anything else.) You can check this by reading the potentiometer as you turn the encoder and looking at how the signals change on the LCD screen. The programs for doing so are in the first half of the book in the Chapter 5.

Names of Registers The PIC 18F4331 uses a number of additional registers to provide the control of all the additional features that are built into the chip. The registers that you are familiar with for use with the 16F877A are not used in the same way even when they have the same or similar names. The same is true for the timers and the A to D control. This means that you must start from scratch when you start to set up this PIC. Your have to read and understand the data sheet for each register.

In the programs that follow, you will notice that a number of registers are set to 0 even though this is how they would normally be set on startup. This is to bring the existence of these important registers to your attention.

Timer Usage The only timer we will use in our programs is TIMER0. You need to get completely familiar with this particular timer in this particular PIC. Like all timers this timer has the following:

- An enable bit
- Global enable consideration
- An on/off bit
- A counter input
- A configuration bit for 8 or 16 bits
- The registers that make up the timer
- Input signal configurations
- Pre-scaler configuration bits and assignments
- An interrupt bit

In general, these are the same things (or similar) to what we were using with the timers in the 16F877A, so the transition should not be difficult. The programs that follow show most of what needs to be done, but you may need to learn more than I have shown for the level of expertise that you seek.

Data Sheet You must download the data sheet for this processor and have it open in a window on your computer as you write your programs.

SETTING UP FOR THE 18F4331

The programmer readings I used for the 18F4331 are shown in Table 10.2. The settings are not rigid, and many variations will work depending on what you want the PIC to do. These options and selections are made in the PIC programmer software.

TABLE 10.2 SETUP INSTRUCTIONS FOR THE PROGRAMMERS OF 18F4331

OSCILLATOR	HS
Internal external switchover	Disabled
Fail safe clock monitor	Disabled
Power up timer	Disabled
Brownout reset	Disabled
Brownout reset voltage	2.0 V
Watchdog Timer	Disabled
Watchdog Timer post-scaler	1:32768
Watchdog Timer window	Enabled
PWM pins	Disabled on reset
Low side transistor polarity	Active High
High side transistor polarity	Active High
Special event reset	Disabled
FLTA input multiplexed with	RC1
SSP I/O multiplexed with	RC 4, 5, 7
PWM 4 multiplexed with	RB5
External clock multiplexed with	RC3
MCLR pin function	Reset
Stack underflow overflow reset	Enabled
Low voltage programming	Enabled
Boot block	Not protected
Codes and all the rest	Not protected
And all the way down	Not protected

11

RUNNING MOTORS: A PRELIMINARY DISCUSSION

There are times when what you need to do with a microcontroller is to move something. The easiest way to do this, of course, is with a small motor or sometimes a solenoid. We will discuss the scope of what we will be covering in the tutorial in this chapter.

Note *Running large motors is very much like running small motors except that you need a much larger amplifier. You also have to provide for more safety interlocks because a lot of energy is being handled, and you have to do everything you can do to keep things from getting out of hand.*

The control of the following types of devices will be covered:

R/C hobby servo motors
Stepper motors
Small DC motors
DC motors with encoders attached for feedback
Relays and solenoids and an AC motor of about one-fourth horsepower

The control of each of these is covered in a section devoted to it.

R/C Hobby Servo Motors

Large servo motors weighing a few pounds or more are available to run off larger power supplies and the signals received from a hobby radio transmitter. The techniques for running these giant servos are the same as those used to run hobby servos used by the hobby radio control industry. As always, safety becomes a serious consideration when we migrate to larger devices and higher currents and voltages.

Of all the motors that we can control with a microcontroller, the easiest to control are the servo motors used by the radio control hobby. These motors have integral gearboxes built into them to allow a movement of about 180 degrees at the output shaft. An internal potentiometer allows the system to determine the position of the output shaft. Control is affected by sending the motor a signal pulse 60 times a second of a required duration of about 1.5. The schemes for providing the control and ideas for using the basic potentiometer feedback for controlling these devices were covered in Chapter 5.

R/C servos can be positioned to one part in 256 if an 8-bit variable is used to control the width of the pulse sent to it and one part in 1024 is a 10-bit variable is used for control. For most purposes served by these small servos, the 8-bit signal is more than adequate for the job, and an 8-bit variable can be read into one byte as compared to the two bytes needed for a 10-bit variable in the MCUs. It takes longer to read the two bytes needed to get all 10-bits read, and this can be important where speed is a prime consideration (which it almost invariably is).

Stepper Motors

Stepper motors move a part of a revolution with each control signal change. They typically contain an arrangement of magnets and coil windings that allow us to move the motor incrementally (in steps). Motors with 400 steps per revolution or 0.9 degrees per step can be obtained at reasonable cost. However, 200 steps per revolution is more common and usually cheaper. If the motor is not allowed to be overloaded (and thus slip), we can keep track of the position of the motor by keeping track of how many times we have sent it a control signal change.

We will cover the control of stepper motors with four wires, or two sets of windings, in detail. These are called bipolar motors and we will use one that needs 12 V at about 1 ampere for our experiments. The amplifiers we use for our servo motors will also be able to control this motor, so no other expense will be involved. Any bipolar motor that has voltages and amperage characteristics that match the amplifiers that we are using can be used.

Other motors are controlled with similar schemes but require more complicated power supplies and electronics. The 4-wire motors require only one power supply, and we can use the amplifiers that we will use for the DC motors to run them. Each amplifier module has to be able to control two sets of motor windings and most "H" bridge type dual amplifiers have this capability.

The two other types of small motors that we are likely to find in the laboratory or hobbyist workbench are small DC motors and DC motors with optical quadrature encoders. In this project we will cover the problems associated with running these types of motors separately and in some detail.

SMALL BRUSH-TYPE DC MOTORS

The control of small or even tiny DC motors can vary from on/off control to simple speed control based on a pulse width modulation (PWM) power signal. The techniques for doing this are covered in Chapter 14. As we try to control larger (but still small) motors, the power needed increases and so does the need for larger electronic controllers to handle the larger power. Integrated circuits that allow us to manage the control of larger motors are now available from a number of manufacturers. We will cover the use of one of these integrated circuits to control the motors in detail (the LMD 18200). A number of suppliers provide readymade amplifiers that use these integrated circuits for controlling small motors. We will cover a number of the amplifiers that are available and discuss how we can use them to control our motors with a microcontroller. The size of the motor we control is limited by the size of the amplifier available to control it. We will limit ourselves to 3 to 6 amperes at between 12 and 55 volts DC provided by the 18200. Since integrated circuits that will handle these amperages and voltages are available at reasonable cost, we will use an amplifier with these characteristics for all our small motors. Three amplifiers we can use are discussed in Chapter 12.

We will not consider brushless motors. Brushless motors are similar to brush motors but use solid state electronics to control the commutator function.

DC Motors with Attached Encoders

In order to control both speed and distance (in revolutions) traveled, you need some sort of feedback mechanism that will tell you how fast the motor is moving and how far the motor has moved. This is usually done with an optical encoder that provides a quadrature signal of a fixed number of cycles per revolution. We count how many cycles have gone by to determine how far the motor has moved and adjust the power fed to the motor as needed for the results we are trying to achieve. The quadrature signal also has the ability to be interpreted for motor direction. The recent availability of microcontrollers that keep track of the encoder counts automatically makes controlling servo motors considerably easier. We will use one of these microcontrollers (the PIC 18F4331) to control an encoder equipped motor. If we have to keep track of the encoder counts, the task becomes much more complicated and is beyond the scope of an essentially minimally technical book like this. In this text I am trying to avoid the use of complicated formulas and assembly language code.

Relays and Solenoids

Though not strictly motors, relays and solenoids use the same magnetic technology that we are using to control the motors to make small movements. These movements are often useful for the experimenter, and we will cover the techniques needed to control these devices without damaging the sensitive electronics in our microcontroller.

SMALL A/C MOTORS AT 120 VOLTS, SINGLE PHASE

Often times it is necessary to control a small AC motor as a part of what we are doing. Controlling the on/off operation of small (under one-fourth hp) is quite straightforward and will be covered in Chapter 17.

Controlling the speed of an AC motor is a little more complicated because these motors are not designed for speed control. Since these motors run at the speed determined by the frequency of the power lines, the easiest way to vary the speed is by varying the frequency of the power line. Even so, there is limit to how much the speed can be changed because of overheating problem in the motor windings and resonances that are related to the overall design of the motor. The sophisticated electronics needed to control the speed of these types of motors are beyond the scope of this book.

"The Response Characteristics" of a Motor

The most important concept that we need to understand is the motor's ability to comply with the command that we send it. If the motor cannot possibly do what we tell it to do, we are essentially wasting time. Trying to control a motor under such circumstances is meaningless. No matter what we tell the motor to do, if the load on the motor, the characteristics of the motor, the motor's power supply, or the controller does not allow the motor to do what we want it to do, no amount of expertise on your part is going to make a difference. Though this seems obvious, it is the reason that most design failures occur.

So what does compliance mean?

In a nutshell, it means that the motor has to have the power to execute the commands sent to it by the controller in real time. Keeping in mind that everything takes time, real time means "right after it gets the command," or "immediately." Usually this becomes a problem when the motor is too weak, the load is too large, or the power supply is inadequate for the task being commanded. The processor we are using has to be up to the task too. It has to have the right features and it has to be fast enough to do the job.

It also means that you have to select a control situation that is realizable if you are going to be successful in your control attempt. It does not mean that we understand the difference between what can and cannot be done at this stage of our learning process. Hopefully, by the time we get to the end of these exercises, we will have a better understanding of what is possible and what is not.

DC MOTOR NOTES

As a general rule, a DC motor needs to be running (under load) at well below 50 percent of the power needed to perform the task at hand. The other 50 percent or so of the power is reserved for the power needed for sudden load changes, to accelerate, and to transition between moves quickly. There will be cases when even more power might be needed. However, keep in mind that a DC motor can put out a lot more than its rated power if the voltage to it is increased. This should be done for short periods only, to prevent overheating.

The limiting condition for a DC motor is the heat that builds up in the motor windings, and the amperage the brushes can transmit to the commutator. If the motor can be kept cool or if the motor materials will handle higher temperatures, DC motors can be pushed well beyond published ratings, especially for short periods of time. We have to monitor the temperature of the motor windings to make sure we do not exceed the temperature the insulation and the wires are designed for. Keep in mind that a little overheating over a long period of time can be just as damaging as a lot of heating in a short time.

The top speed of the motor is limited by the back EMF that it generates as its speed increases. When the back EMF gets high enough, the motor can no longer increase its speed. At higher amperages the ability of the brushes to transmit the necessary current to the commutator is compromised, and sparking at the brushes increases. This sparking destroys the commutator and the brushes very quickly.

For our purposes we can consider the motor's response to the voltage applied to be linear. This means, in general, twice the voltage will give us twice the speed. We will assume that we will be working within electrical parameters that the motors will tolerate without difficulty.

12

MOTOR AMPLIFIERS

Before we start our discussion on motor control, let us take a look at a few small amplifiers that we can use to control our motors. All the amplifiers I have selected are inexpensive and easy to use. Other suppliers provide similar amplifiers you might find more suitable for your particular application, but I did not investigate any of them, and the circuits provided in this tutorial do not cover the use of any other amplifiers. However, you should not have any trouble with using the other amplifiers. If you want to run a larger motor, all you need is a larger amplifier.

The three amplifiers shown in Figure 12.1 provide an inexpensive way to run the motors we are interested in. We need a 2-axis amplifier to run the stepper motors, so if you are going to buy only one amplifier buy one of the two 2-axis amplifiers. The Solarbotics amplifier is cheaper, but it also handles fewer amps, as shown in Table 12.1. The amplifier I used for all the experiments in this book is the Xavien 2-axis amplifier on the right in Figure 12.1.

The three amplifiers take TTL signals directly from the microcontroller and control the power to the motor. Each motor requires a power supply that matches the power needed by the motor and the capacity of the amplifier. The power supply of the microcontroller and the power supply of the motor should be kept separate under all circumstances with only a common ground connection. If this is not done, noise from the motors will contaminate the power to the microprocessor and cause severe problems. All motors are very noisy as far as computer electronics are concerned and must be isolated. Motor noise comes from the motor commutators and from the rapid on and off switching of the motor coils. Though the addition of small capacitors to ground from each motor terminal and across the terminals helps, it does not work as well as a well isolated layout. Since we have a choice, we will use separate power supplies in all our experiments.

Each of the amplifiers uses one or two integrated circuits as its amplifier components, and some provide ancillary LEDs to annunciate internal conditions. Still other

TABLE 12.1 BASIC AMPLIFIER PROPERTIES

AMPLIFIER	CHIP USED	MAX. VOLTAGE	AMPERAGE	COMMENTS
Xavien 1 axis	33886	40	5	Freescale semi
Solarbotics 2-axis	L298	46	2 each	
Xavien 2-axis	LMD18200	55	3 each	Amplifier I used

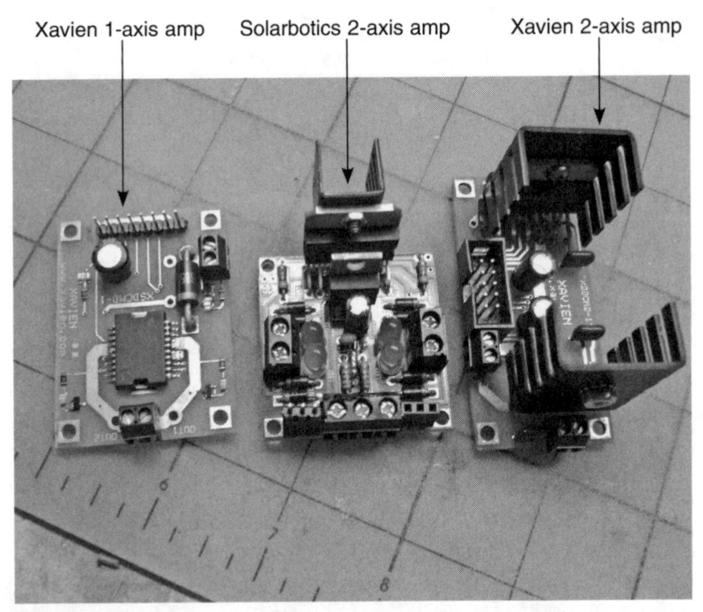

Figure 12.1 **Small, inexpensive amplifiers for small running motors.**

devices allow interfacing to the signals that the microcontroller provides without the need for any intermediate devices.

The capacities of the amplifiers are listed in Table 12.1.

Notes on Homemade Amplifier Construction

Though you can make your own amplifier, I do not recommend that you do this other than as an interesting exercise. The amplifiers you are likely to make (a number of designs are available on the Internet) are likely to be fairly straightforward H bridges. Unless considerably more sophisticated circuitry is added to the basic amplifier circuit, it is very easy to blow up an H bridge by turning on both transistors on any one side of the bridge on at the same time. Trust me: homebrew amplifiers are unbelievably easy to destroy.

On the other hand, if we use purchased integrated circuits to build our amplifiers, these circuits will almost certainly have circuitry within them to prevent damage on short circuiting, shutdown on overheating, and other useful features. The more sophisticated circuits also provide the ability to detect thermal shutdown and to look at the current flow through each amplifier.

Since there are a number of vendors that are willing to sell you inexpensive, ready to use amplifiers, there is no good reason at this stage in our learning process to not use these resources to run our motors. All discussions and circuits in the tutorial will reflect this.

Reduce the stress in your life. Buy an amplifier.

The Xavien 2-Axis Amplifier

The amplifier I used for all the experiments in this book is the Xavien 2-axis amplifier shown in Figure 12.2. The connections that this amplifier used are identifies in Figure 12.3. Each of the two amps on the Xavien can handle up to 3 amps at 55 VDC. Short pulses of 6 amps are tolerated.

The polarity of the power to the amplifier is critical. *It must be observed.* No other protection is provided.

Table 12.3 lists the pin functions for the ten control lines in a Xavien amplifier.

We will not be using the current sense and thermal flag lines. They add nothing to the ideas about running motors and can be ignored at this stage.

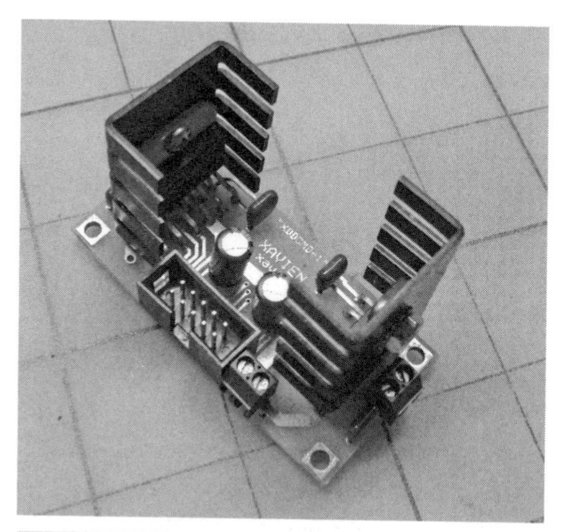

Figure 12.2 The Xavien 2-axis amplifier

TABLE 12.2 XAVIEN 2-AXIS AMPLIFIER PIN FUNCTIONS

PIN	FUNCTION
1	Motor 1 brake
2	Motor 1 PWM or enable/run
3	Motor 1 direction
4	Motor 1 current sense, analog (not used in the discussions)
5	Motor 1 thermal flag, digital (not used in the discussions)
6	Motor 2 direction
7	Motor 2 brake
8	Motor 2 PWM or enable/run
9	Motor 2 thermal flag, analog (not used in the discussions)
10	Motor 2 current sense, digital (not used in the discussions)

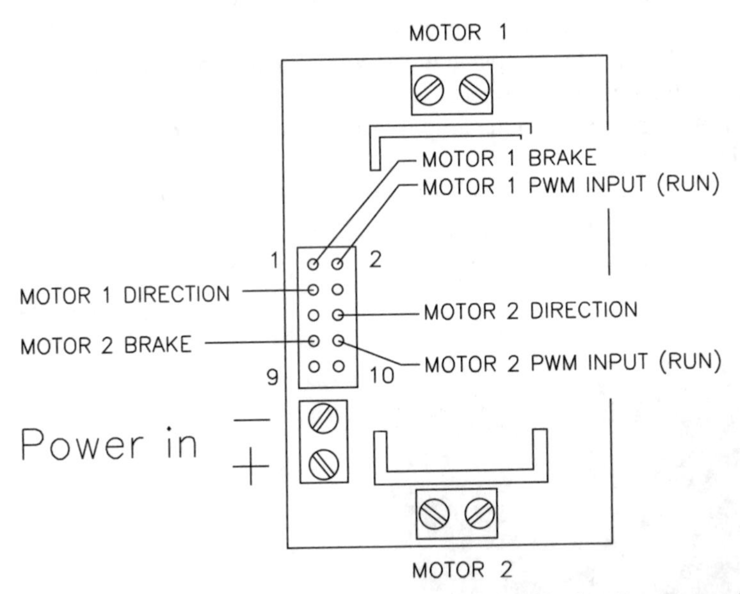

Figure 12.3 **Connection schematic for the Xavien 2-axis amplifier**

However, we *cannot* ignore the brake lines because they have to be pulled to zero either in software or with hardware to turn the brake off.

For our purposes pins 1, 2, and 3 can be used to control coil/motor 1, and pins 6, 7, and 8 will control coil/motor 2. One way of wiring a DC motor and the LAB-X1 board to do this using one half of the Xavien amplifier is shown in Figure 12.4.

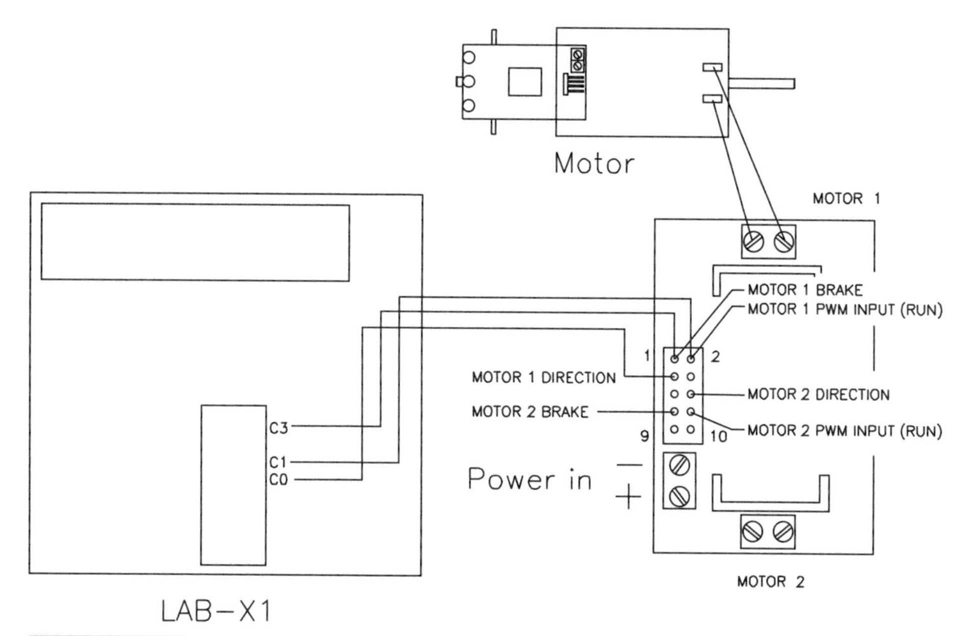

Figure 12.4 Using the Xavien 2-axis amplifier: one possible scheme for connecting a motor to one of the amplifiers

The 1-Axis Xavien Amplifier

If you need a single axis amplifier, the single axis Xavien amplifier shown in Figure 12.5 is suitable for small motors needing less than 5 amps at 40 V. I used this amplifier for the small DC motor experiments. The wiring connections for this amplifier are shown in

Figure 12.5 The Xavien 1-axis amplifier

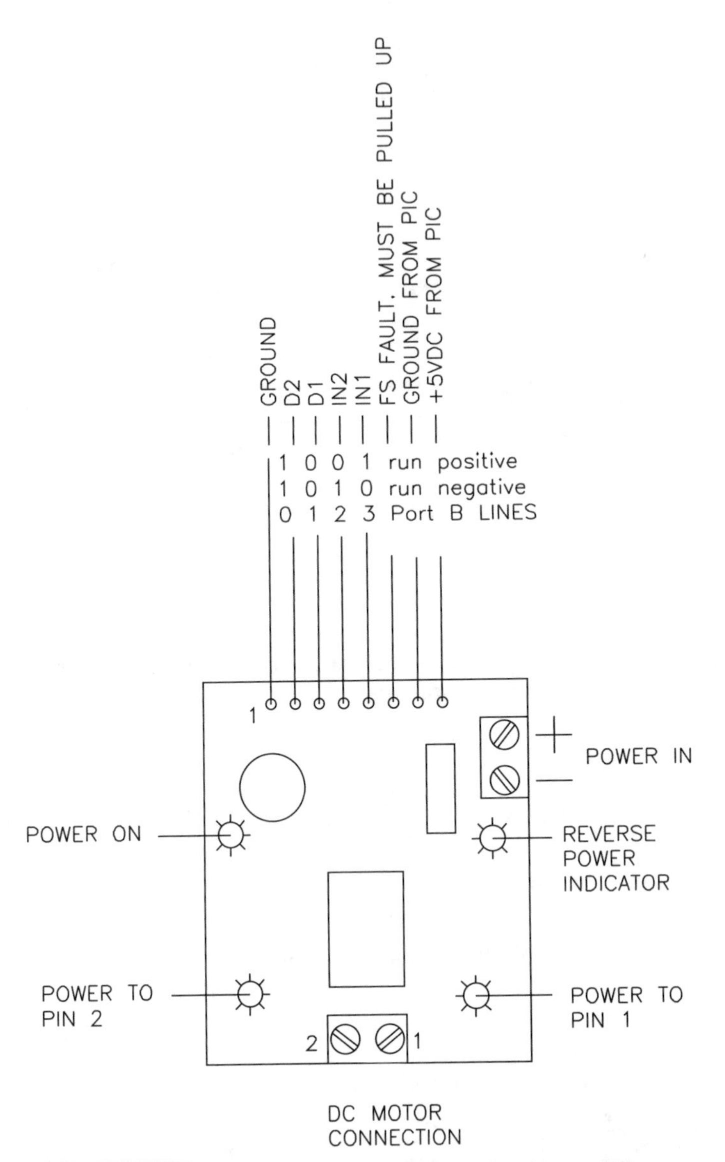

Figure 12.6 Schematic of the Xavien 1-axis amplifier using PORTB

Figure 12.6 and described in Table 12.3. A safety feature on this amplifier provides a diode to protect against accidental reverse polarity connection to the power connector.

Table 12.3 provides a key to the eight lines controlling the amplifier; refer also to Figure 12.7.

- **Power in** The power for the motor connects to these two terminals. Though a protective diode is provided, the polarity of the connection should be observed.

TABLE 12.3

LINE	LABEL	DESCRIPTION
1	Ground	Common ground
2	D2	Disable; active high input to tri-state outputs
3	D1	Disable; active high input to tri-state outputs
4	IN2	Logic input control of OUT2
5	IN1	Logic input control of OUT1
6	FS fault	Fault sense for H bridge; open drain active low output; fault line must be pulled up with 10K to 100K
7	Ground	Ground
8	5VDC	Logic power

- **DC motor** This is where the motor is connected. The polarity of this connection is not important. The motor operations can be reversed in software.
- **LEDs** The three LEDs on the card indicate the operation of the card as power and control signals are applied to the card.

An example of a circuit for controlling the single-axis Xavien amplifier with a PIC16F877A is provided in Figure 12.7.

No programs are given for this specific amp in this book. It can be used for all single-axis work.

The Solarbotics 2-Axis Amplifier

The 2-axis Solarbotics amplifier is shown in Figure 12.8 and its wiring connections are identified in Figure12.9. The Solarbotics amplifier is provided as a kit, and the kit is easy to solder together. An example of a circuit for controlling the Solarbotics amplifier with a PIC16F877A is shown in Figure 12.10. This circuitry would allow the control of two motors.

Note *No programs are given for this amp in this book. This is an inexpensive kit recommended for those on a tight budget. It is best for small loads.*

The Solarbotics amplifier has a problem with PWM signals under certain conditions and should not be used for sophisticated experiments with a lot of PWM changes on both axes simultaneously. There are some crosstalk and noncompliance issues.

The condition does *not* inhibit the use of this inexpensive amplifier for the single-axis operation of small motors. This is about the cheapest integrated 2-axis amplifier that you can buy, so you may want to consider it.

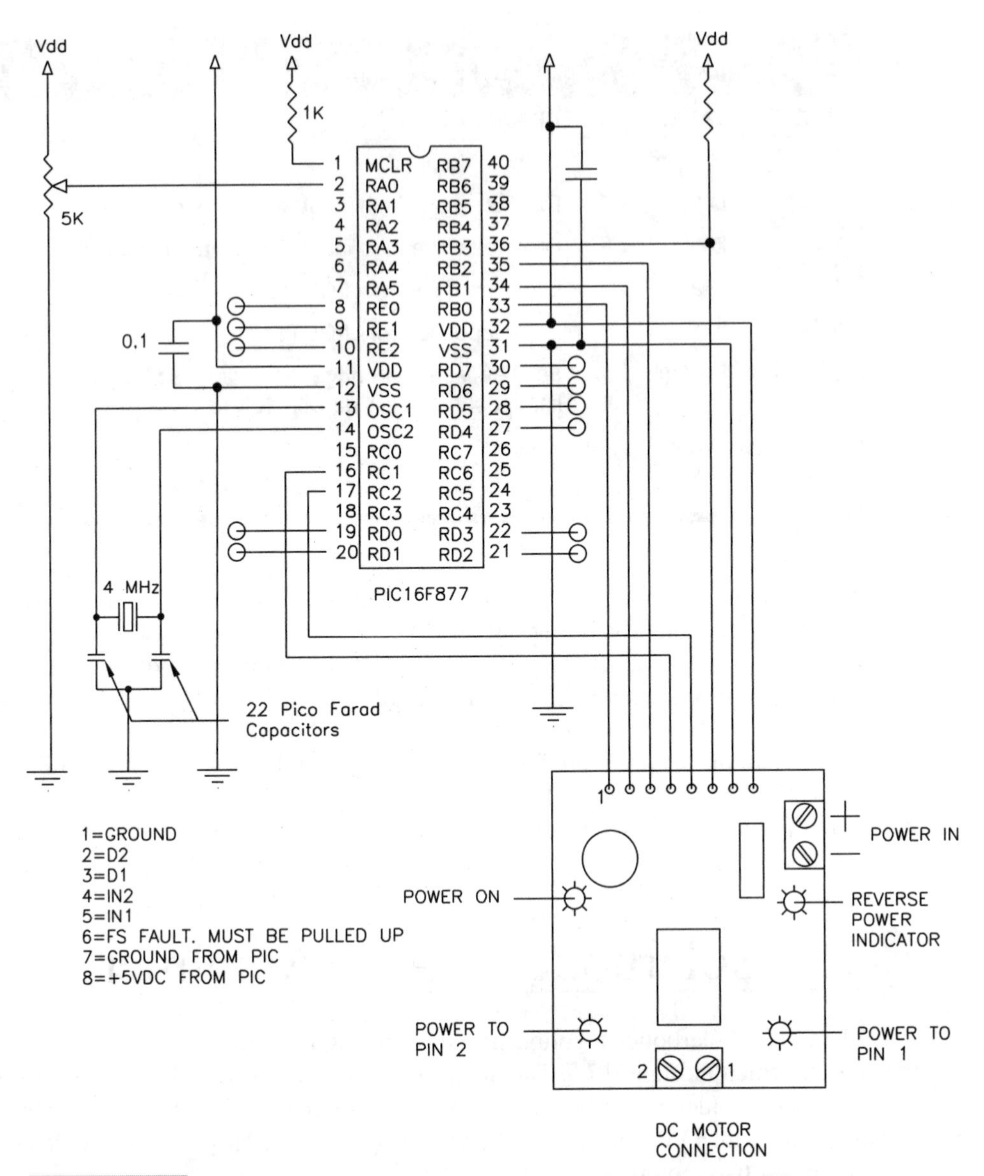

Figure 12.7 One way of controlling the single-axis Xavien amplifier

Figure 12.8 The 2-axis Solarbotics amplifier

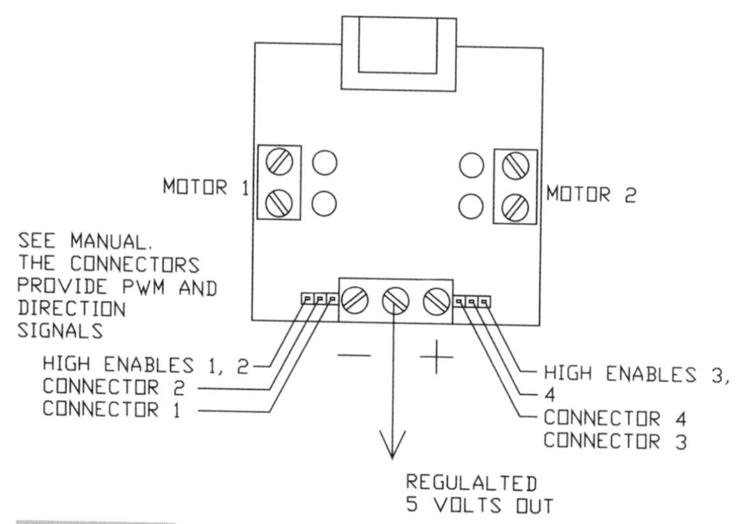

Figure 12.9 The 2-axis Solarbotics amplifier connections

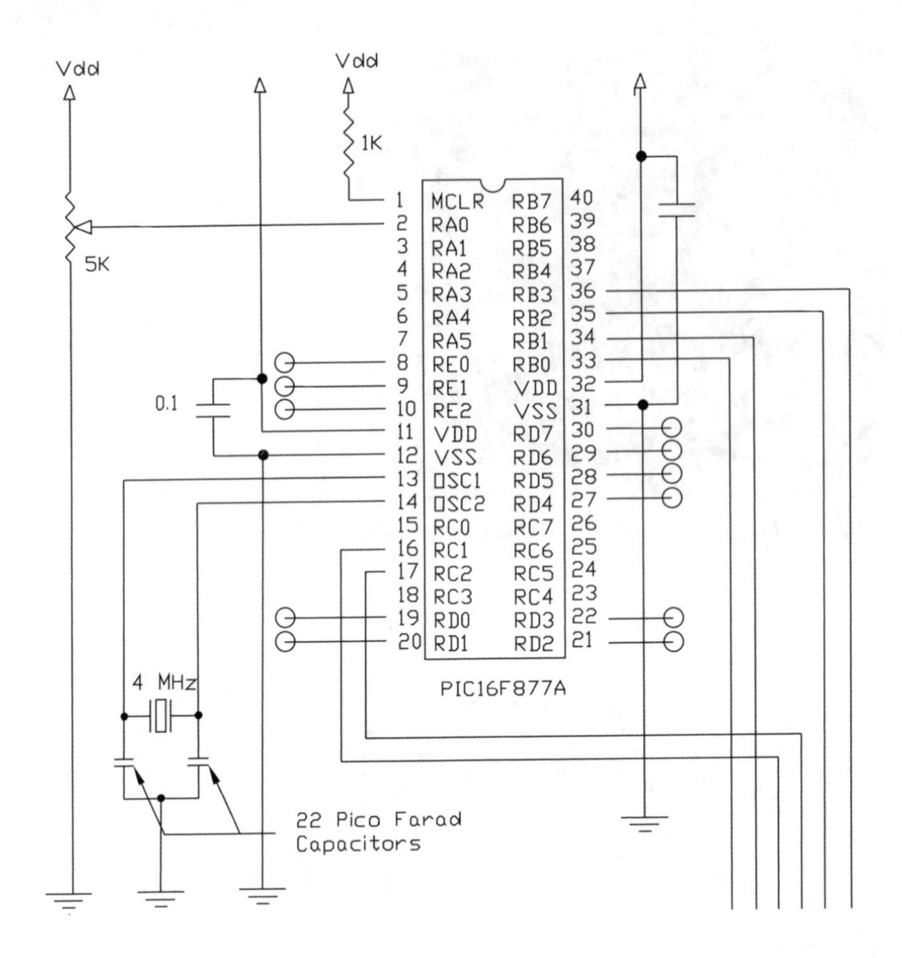

CONNECTIONS TO THE AMPLIFIER ARE DETERMINED BY THE CONTROL DESIRED. EACH SIDE OF THE AMPLIFIER PROVIDES THREE CONNECTIONS: ENABLE, PWM, AND REVERSE.

SEE SOLARBOTICS MANUAL

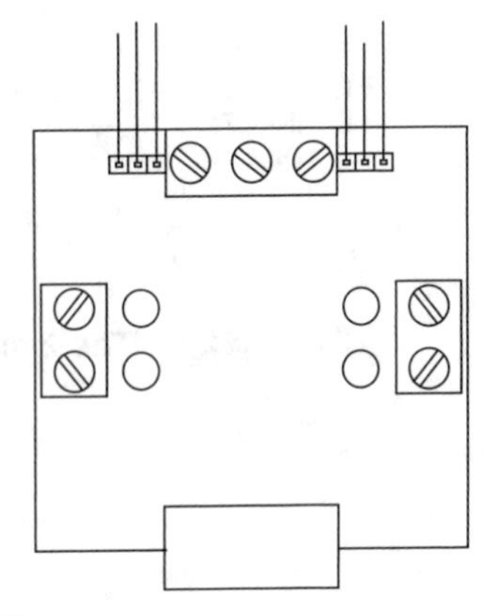

Figure 12.10 Using the Solarbotics 2-axis amplifier

13

RUNNING HOBBY R/C SERVO MOTORS

Of all the different types of motors that we can control with a microcontroller, the easiest to control motors are the servo motors provided for the model aircraft radio control hobby. These small motors use a three wire interface consisting of power, ground and the control signal. They do not need an amplifier of any sort! See Figure 13.1.

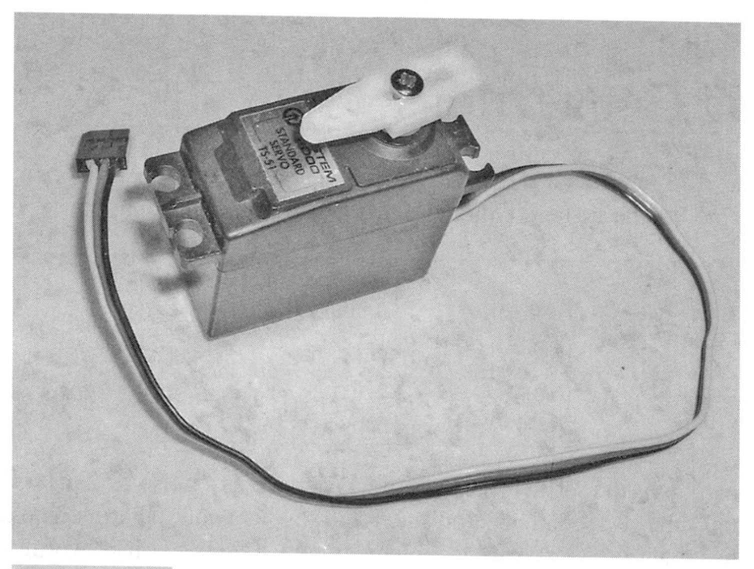

Figure 13.1 A typical model aircraft R/C servo that uses standard Futaba wiring

Model Aircraft Servos

Before we start, a special requirement for model aircraft servos must be noted: these servos need to have a pulsed signal sent to them approximately 60 times a second on a regular basis to accurately maintain their commanded position. If this is not done, the operation of the motors becomes jerky and irregular. This one-sixtieth of a second requirement also means that there can be a minimum worst case lag of about a sixtieth of a second whenever a command is sent to an R/C servo. For all practical purposes, a lag this long is not critical in a motor application; however, you need to keep in mind that this delay does exist. This also means that it takes one-sixtieth of a second (worst case) to respond to a command.

Now that we know that the servos have to be reminded of their position about 60 times a second in order to maintain proper operation, we need a way to pulse the servos on a regular basis of 60 times a second. Since typical program flow timing is indeterminate, the pulses cannot effectively be made a part of the standard program flow and still guarantee that the servos will get pulsed as needed. Another scheme is needed.

To guarantee that the motors will get a signal 60 times a second we will need to set up an interrupt routine that will be called 60 times a second with one of the timers. The interrupt routine refreshes the counters in the servo pulse generator and sends out the necessary pulse every time it is called. The length of the pulse itself needed is determined within the program as it runs it course. As the calculated positions for the application we have in mind are determined, they are sent to the motors. An interrupt-driven system is one way of guaranteeing the smooth operation of these servos.

Wiring Connections

The standard R/C servo is a three-wire device. The signals on the three wires are as follows for the Futaba system. Other systems may vary but are similar.

Power	Red	Specified by the manufacturer (usually 5 V will work)
Ground	Black	Ground
Control signal	White	A TTL level signal—this is the pulsed signal connection

In Futaba systems the servo center position is defined as a pulse 1.52 ms wide delivered about 60 times a second. The pulse width range is about 0.75 ms on either side of that. Other manufacturers specify values around 1.5 ms, so it is worth to check exactly what your servos need in the way of the center positioning signal and the range. Also check to see that the wiring matches what you are going to provide with your MCU.

Fairly large servos that follow schemes similarly to the R/C operation standard are made for industry, though they are probably beyond affordability for most students and hobbyists. These servos can provide adequate power for demanding laboratory and industrial applications.

DETERMINING THE POSITIONS OF THE SERVO

When we put a servo to use, it will have to move to certain specific positions to do the work we need done. We need a way to determine the exact positions needed in our applications for each servo so that we can set the positioning parameters to the appropriate values in our programs. The program that we are about to create will allow you to move both servos under computer control from potentiometer 0 and 1 and watch the signal values that are being sent to the servos on the LCD (see Program 13.1). With this program, we can find with ease the positional values for each of the servos we will be using whenever we need to. The program uses the two lower potentiometers on the LAB-X1 to control the positions of the servo in real time. Adjust the potentiometers as needed to get the limit position for each servo and then put these values into your program. POT0 determines the position of the servo and POT1 determines the delay between the movements as the servo moves back and forth.

Program 13.1 This is a "stand alone" program for finding the exact servo setting to determine the position of a servo. No interrupts are being used in the program.

```
CLEAR                           ; start with clearing the memory
DEFINE OSC 4                    ; define the oscillator
DEFINE LCD_DREG PORTD           ; define lcd connections
DEFINE LCD_DBIT 4               ; 4 bit protocol
DEFINE LCD_RSREG PORTE          ; register select byte
DEFINE LCD_RSBIT 0              ; register select it
DEFINE LCD_EREG PORTE           ; enable port
DEFINE LCD_EBIT 1               ; enable bit
LOW PORTE.2                     ; leave low for write
ADCON1 = %00000111              ; set a to d control register
PAUSE 500                       ; pause for 0.5 seconds for lcd startup
TRISC = %00000000               ; PORTC is all outputs for servos
PORTC = %00000000               ; set to 0
PORTD = %00000000               , PORTD is all outputs for LCD
PORTE = %00000000               , PORTE is all outputs for LCD
LCDOUT $FE, 1, "CLEAR"          ; clear display and show CLEAR
DEFINE ADC_BITS 8               ; set number of bits in result
DEFINE ADC_CLOCK 3              ; set internal clock source (3=rc)
DEFINE ADC_SAMPLEUS 150         ; set sampling time in µs
POT0 VAR BYTE                   ; create adval to store result
POT1 VAR BYTE                   ; create adval to store result
```

(continued)

Program 13.1 This is a "stand alone" program for finding the exact servo setting to determine the position of a servo. No interrupts are being used in the program. (*continued*)

```
LOOP:                          ; main loop
  LCDOUT $FE, $80, DEC3 POT0," ", DEC3 POT1," " ; print data
  ADCIN 0, POT0                ; read port a0
  ADCIN 1, POT1                ; read port a1
  POT0 = POT0 + 23             ; so actual pulse is displayed
  PULSOUT PORTC.1, POT0        ; pulse port C.1
  PAUSE POT1                   ; pause 1/60 seconds approx
  PULSOUT PORTC.0, POT0        ; pulse port C.2
  PAUSE POT1                   ; pause 1/60 seconds, approx value
                               ; is 24
GOTO LOOP                      ; do it again
END                            ; always end with END
```

In Program 13.1, the pulses to both connectors J7 and J8 are provided with the same value, POT0, in milliseconds. Doing it this way allows us to connect a servo up to either connector and the operation will be identical. Or, both servos can be connected and you can find the end positions for both connected servos with this one program. Both servos will not have the same value for the positive and negative positions for their respective mechanisms—unless of course, you can make the linkages absolutely identical and the servos themselves are guaranteed to be mechanically and electrically identical. (As previously mentioned, POT1 controls the delay between the pulses. Experiment with this delay to see what happens to the operation of the servos as you change the pause time between signals.)

ADDING THE INTERRUPT ROUTINE

Next, let us set the conditions needed to update the servo positions from an interrupt routine that gets called every sixtieth of a second. We will use TIMER0 to do this, but any timer could be used. Remember that TIMER1 is the default timer for the HPWM generator. If we wanted to use TIMER1, we would have to specify a different timer for the HPWM command. See the PICBASIC PRO manual for details.

Now that we know we are on the right path, let us create a program to toggle D.0 for a 1 second on and 1 second off time based on an interrupt routine. This is done in Program 13.2. Once we get this working, we will modify the program to provide the 60 interrupts a second that we need.

Program 13.2 One second blinker on PORTD.0

```
CLEAR                          ; start with clearing the memory
DEFINE OSC 4                   ; define the oscillator
```

(*continued*)

Program 13.2 One second blinker on PORTD.0 *(continued)*

```
DEFINE LCD_DREG PORTD   ; define lcd connections
DEFINE LCD_DBIT 4       ; 4 bit protocol
DEFINE LCD_RSREG PORTE  ; register select byte
DEFINE LCD_RSBIT 0      ; register select it
DEFINE LCD_EREG PORTE   ; enable port
DEFINE LCD_EBIT 1       ; enable bit
LOW PORTE.2             ; leave low for write
ADCON1 = %00000111      ; sets the A to D control register
PAUSE 500               ; pause for 0.5 seconds for LCD startup
                        ;
LCDOUT $FE, 1, "One second blinker" ; clear display and
                                    ; show title
ON INTERRUPT GOTO INT_ROUTINE   ; target interrupt routing
OPTION_REG = %00000111  ; sets pre-scalers etc
INTCON = %00100000      ; sets interrupt
X VAR BYTE              ; Define variables
Y VAR BYTE              ;
Y = 0                   ;
X = 0                   ; initialize variables
                        ;
LOOP:                   ; main loop
LCDOUT $FE, $C0, BIN8 X ," ",DEC2 Y  ; update the display
PAUSE 10                ; so you can see the display
GOTO LOOP               ; do it again
                        ;
DISABLE                 ; disable and enable bracket the
                        ; interrupt routine
INT_ROUTINE:            ; the interrupt routine
  X = X+1               ;
  IF X = 200 THEN       ; goes through x loop 200 times
    Y = Y+1             ; before incrementing Y
    X = 0               ; and resetting X
  ELSE                  ;
  ENDIF                 ;
  IF Y = 2 THEN         ; Checks for Y and resets it if
                        ; it is 2
    Y = 0               ; resets Y
    TOGGLE PORTD.0      ; flips state of LED on D.0
  ELSE                  ;
  ENDIF                 ;
RESUME                  ; go back to program
ENABLE                  ; disable and enable bracket the
                        ; interrupt routine
                        ;
END                     ; end program as usual
```

In Program 13.2, play around with the prescaler embodied in OPTION_REG and the counts in the variables *X* and *Y*. Together these determine the rate at which D.0 blinks. As written in Program 13.2, the blink cycle is once every 2 seconds. Note that the rate is determined approximately by the product of *X* and *Y*. We are using 1 second at this stage because a blink 60 times a second is impossible to see with the human eye (but could be seen easily on an oscilloscope).

Looking at it another way, Program 13.3 is a program that can be used to determine how long the pulses to control the servos need to be without any interrupts or fancy footwork. This is not the recommended way to go about this but it shows you a way of getting the pulses generated. This program would have problems it other tasks had to be undertaken as the servos were controlled. Glitches would appear in the operation of the servos.

Program 13.3 **Simple servo position determination.**

```
CLEAR                        ; clear memory
DEFINE OSC 4                 ; define the oscillator
DEFINE LCD_DREG PORTD        ; define lcd connections
DEFINE LCD_DBIT 4            ;
DEFINE LCD_RSREG PORTE       ;
DEFINE LCD_RSREG PORTE       ;
DEFINE LCD_RSBIT 0           ;
DEFINE LCD_EREG PORTE        ;
DEFINE LCD_EBIT 1            ;
LOW PORTE.2                  ; LCD R/W line low (w)
ADCON1 = %00000010           ; PORTE to digital.
PAUSE 500                    ; wait 0.5 seconds for LCD startup
TRISC = %00000000            ; make PORTC all outputs
PORTC = 0                    ; turn off all pins on PORTC
X VAR WORD                   ; Set the variables
Y VAR WORD                   ;
AD1 VAR BYTE                 ;
Y = 150                      ; initialize Y
DEFINE   ADC_BITS 8          ; set number of bits in result
DEFINE   ADC_CLOCK 3         ; set internal clock source (3=rc)
DEFINE   ADC_SAMPLEUS 50     ; set sampling time in µs
LCDOUT $FE, 1                ; clear the display
                             ;
LOOP:                        ;
   ADCIN 0, AD1              ; read the pot
   LCDOUT $FE, $80, DEC5 (200+(8 * AD1))   ;
   HIGH PORTC.1              ; make PORTC.1 high
   PAUSEUS (200+(8 * AD1))   ; make the pause
   LOW PORTC.1               ; make PORTC.1 low
   PAUSE 10                  ; pause to see display
GOTO LOOP                    ; continue to loop
                             ;
END                          ; end the program as usual
```

Next, let us combine the two programs and add the code needed to specify and update the pulses being sent to a servo. This will mean adding to the main loop and adding the interrupt routine. Our goal is to pulse the servo approximately 60 times a second based on an interrupt.

Putting in the missing statements gives us Program 13.4.

Program 13.4 Servo control program with interrupts.

```
CLEAR                               ; start with clearing the
                                    ; variables
DEFINE OSC 4                        ; define the oscillator
DEFINE LCD_DREG PORTD               ; define lcd connections
DEFINE LCD_DBIT 4                   ; 4 bit protocol
DEFINE LCD_RSREG PORTE              ; register select byte
DEFINE LCD_RSBIT 0                  ; register select it
DEFINE LCD_EREG PORTE               ; enable port
DEFINE LCD_EBIT 1                   ; enable bit
LOW PORTE.2                         ; leave low for write
ADCON1 = %00000111                  ; set PortE to digital
PAUSE 500                           ; pause for 0.5 seconds for
                                    ; lcd startup
                                    ;
LCDOUT $FE, 1, "SERVO LIMITS"       ; clear display and show clear
PAUSE 500                           ;
LCDOUT $FE, 1                       ; clear the display again
ON INTERRUPT GOTO INT_ROUTINE       ; target for the interrupt
                                    ; routing
OPTION_REG = %00000011              ; set the prescaler
INTCON = %00100000                  ; enable to interrupt flag
                                    ;
POT1 VAR BYTE                       ; variable created
POT2 VAR BYTE                       ; variable created
POS VAR WORD                        ; variable created
X VAR BYTE                          ; variable created
Y VAR WORD                          ; variable created
Y = 0                               ; set variable
X = 0                               ; set variable
                                    ;
LOOP:                               ; main loop
  ADCIN 0, POT1                     ; read POT1
  ADCIN 1, POT2                     ; read POT2
  POS = POT1*8+(POT2/32)            ; do calculation for POS
  Y = Y+1                           ; increment y in foreground
                                    ; task
  LCDOUT $FE, $80, "Adjustments =", DEC4 POS," ", DEC1
(POT2/32)," "                       ; display
  LCDOUT $FE, $C0, "X=",DEC3 X ," Y=",DEC5 Y  ; display items
                                              ; of interest
```

(continued)

Program 13.4 Servo control program with interrupts. (*continued*)

```
GOTO LOOP                       ; do it again
                                ;
DISABLE                         ; disable interrupts
INT_ROUTINE:                    ; interrupt routine
  X = X+1                       ; increment x
  IF X = 5 THEN                 ; check value of x
    X = 0                       ; reset x
    PORTC.1 = 1                 ; make PORTC.1 high
    PAUSEUS POS                 ; pause to match position
                                ; of servo
    TOGGLE PORTC.1              ; make PORTC.1 low again
  ELSE                          ; logic
  ENDIF                         ; logic
  INTCON.2 = 0                  ; reset/clear the interrupt
                                ; flag
RESUME                          ; resume main program
ENABLE                          ; enable interrupts again
END                             ; all programs must end
                                ; with end.
```

In Program 13.4, the pulse length is set with POT1 for coarse control and POT2 for fine control. That is why POT2 is divided by 8 before adding to POT1. The foreground task in Program 13.4 is counting into the *Y* register, and the background task is controlling the servo.

Now that we know how to control the servo position with an interrupt driven routine in the background, we can write a program to cycle the servo where the limits of the positions are controlled by POT1 and POT2. The limits need to be between 0 and 2400. We will have to multiply the maximum potentiometers reading of 255 by 10 to cover the range.

Adding the servo movement constants to the foreground loop, we get Program 13.5.

Program 13.5 Finding Servo limits (with interrupt driven update timing).

```
CLEAR                           ; start with clearing the
                                ; variables
DEFINE OSC 4                    ; define the oscillator
DEFINE LCD_DREG PORTD           ; define lcd connections
DEFINE LCD_DBIT 4               ; 4 bit protocol
DEFINE LCD_RSREG PORTE          ; register select byte
DEFINE LCD_RSBIT 0              ; register select it
DEFINE LCD_EREG PORTE           ; enable port
DEFINE LCD_EBIT 1               ; enable bit
LOW PORTE.2                     ; leave low for write
ADCON1 = %00000111              ; set PortE to digital
```

(*continued*)

Program 13.5 Finding Servo limits (with interrupt driven update timing). (*continued*)

```
PAUSE 500                            ; pause for 0.5 seconds for
                                     ; lcd startup
                                     ;
LCDOUT $FE, 1, "SERVO LIMITS"        ; clear display and show clear
PAUSE 500                            ;
LCDOUT $FE, 1                        ; clear the display again
ON INTERRUPT GOTO INT_ROUTINE        ; target for the interrupt
                                     ; routing
OPTION_REG = %00000011               ; set the prescaler
INTCON = %00100000                   ; enable to interrupt flag
                                     ;
POT1 VAR BYTE                        ; variable created
POT2 VAR BYTE                        ; variable created
POS VAR WORD                         ; variable created
X VAR BYTE                           ; variable created
Y VAR WORD                           ; variable created
Y = 0                                ; set variable
X = 0                                ; set variable
                                     ;
LOOP:                                ; main loop
  ADCIN 0, POT1                      ; read POT1
  ADCIN 1, POT2                      ; read POT2
  POS = POT1*10                      ; do calculation for POS
  FOR Y = 1 TO 100                   ; ]
    PAUSE 10                         ; ] delay loop
  NEXT Y                             ; ]
  LCDOUT $FE, $80, DEC4 POS          ; display items
  POS = POT2*10                      ; ]
  FOR Y = 1 TO 100                   ; ] delay loop
    PAUSE 10                         ; ]
  NEXT Y                             ; increment y in foreground
                                     ; task
  LCDOUT $FE, $C0, DEC4 POS          ; display items of interest
GOTO LOOP                            ; do it again
                                     ;
DISABLE                              ; disable interrupts
INT_ROUTINE:                         ; interrupt routine
  X = X+1                            ; increment x
  IF X = 5 THEN                      ; check value of x
    X = 0                            ; reset x
    PORTC.1 = 1                      ; make PORTC.1 high
    PAUSEUS POS                      ; pause to match position
                                     ; of servo
    TOGGLE PORTC.1                   ; make PORTC.1 low again
```

(*continued*)

Program 13.5 **Finding Servo limits (with interrupt driven update timing).**
(*continued*)

```
        ELSE                            ; logic
        ENDIF                           ; logic
        INTCON.2 = 0                    ; reset/clear the interrupt
                                        ; flag
    RESUME                              ; resume main program
    ENABLE                              ; enable interrupts again
    END                                 ; all programs must end
                                        ; with end.
```

In Program 13.5, whatever goes on in the LOOP does not bother the execution of the commands to update the servo positions because the servo subroutine is interrupt driven. This is the basic technique for handling all timing-critical tasks. You must get familiar with using this technique and be completely comfortable with it. This technique will also be used to space servo motor stepping commands for the constant/even speed control that they need for speed control.

In the previous programs we converted the position of a potentiometer to a position of the R/C servo. In our particular case the potentiometer was read as a value between 0 and 255, and the servo arm moved approximately 180 degrees. For most applications, only about 90 degrees of movement of the servo is useful. This being the case, we might want to modify the software so that the entire 256 values read from the potentiometer are mapped to the 90 degree movement. This will also give us a finer control of the movement.

THE LAB-X1 CIRCUITRY USED TO CONTROL THE SERVOS

Figure 13.2 contains the relevant circuitry for running the two R/C servos from a 16F877A. All this circuitry exists, as shown, on a LAB-X1 board. The rest of the circuitry is suppressed to reduce the confusion.

The LAB-X1 can be used without modification to run R/C servomotors.

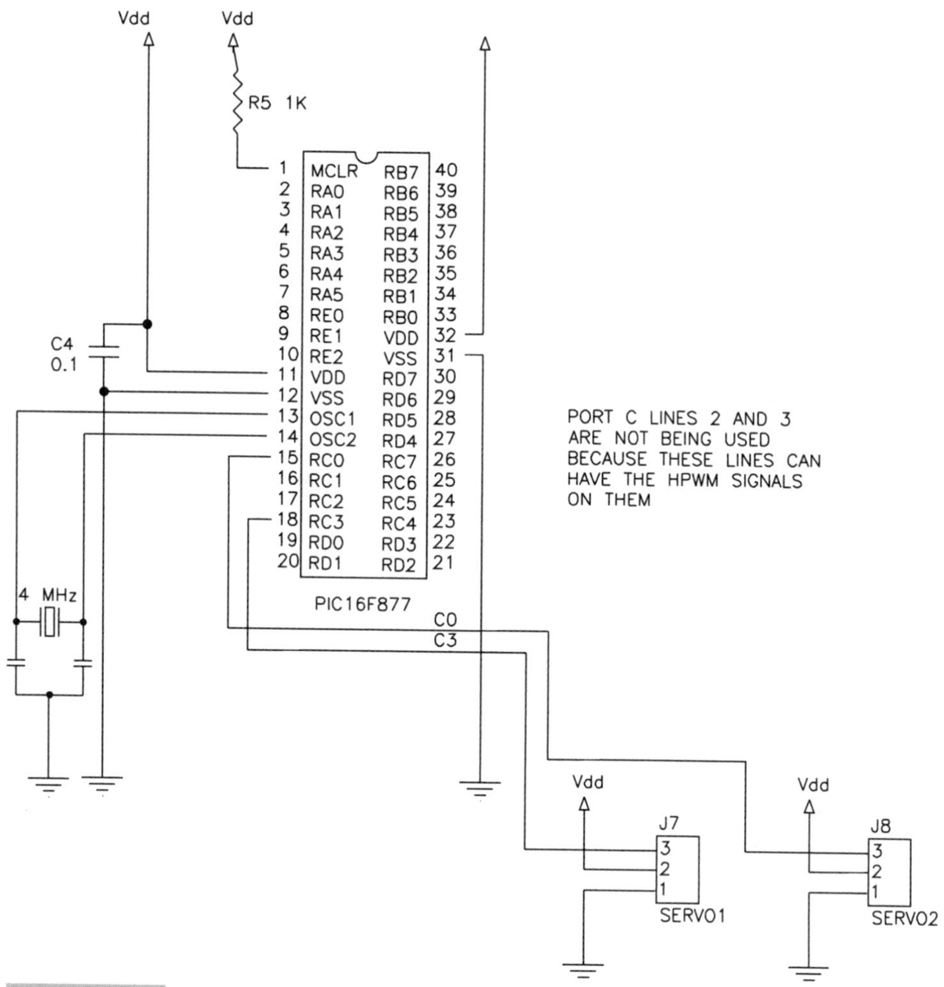

Figure 13.2 Wiring diagram: circuitry for two servos run from a 16F877A microcontroller

14

RUNNING SMALL DC MOTORS WITH PERMANENT MAGNET FIELDS

For our immediate purposes, let us define small DC motors as those about an inch or two in diameter and two to four inches long. The types of motors we will be considering are shown in Figure 14.1. We will use ones that run on 6 to 24 V and draw a couple of amps. The amperage and voltage values have to match the capacity of the amplifiers we have chosen for running the motors. (The 2-axis Xavien amplifier needs a minimum of 12 V to operate properly and will handle 3 amps continuously and 6 amps for short bursts at up to 55 VDC.

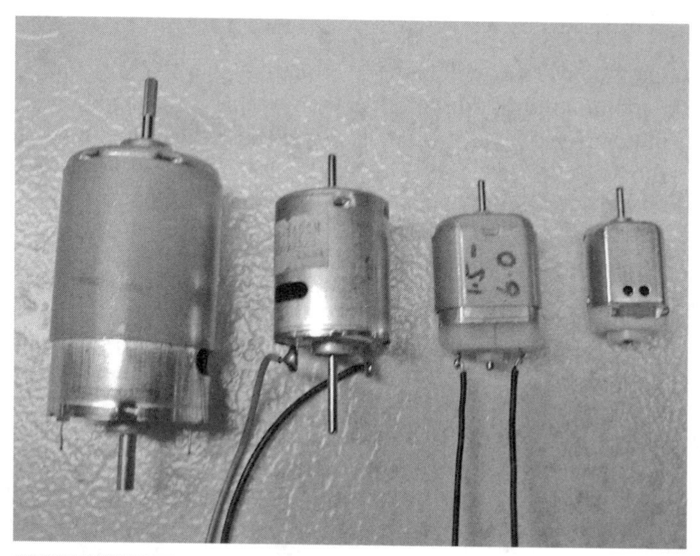

Figure 14.1 Examples of small DC motors under discussion. Motors with shafts on both ends will allow us to mount an encoder directly on one end for later experiments.

Like all DC motors, the small motors shown in Figure 14.1 provide high speed and low torque. They provide no feedback regarding the distance traveled (revolutions completed) or the speed of the motor. (Under certain conditions the back EMF generated by a motor can be used as speed feedback, but we will not use this in our experiments. This use is more common under analog control schemes.) In later chapters we will add encoders to these motors to provide the *digital* speed and distance feedback we need for a more comprehensive control of these motors.

On these simple DC motors we can control the following parameters:

- On/off control
- %Power to motor
- Polarity of power provided (direction of movement)
- Minimum power delivered at starting set point (power needed to start motor)
- Maximum power delivered when running as a set point (maximum rpm control, depends on load)

Essentially, we can have comprehensive control of both the speed and the direction of these motors. Let us design a system that will give us this control of the motor from a potentiometer on the LAB-X1 with the Xavien 2-axis amplifier. The middle position of the potentiometer will be the zero speed position. As we turn the potentiometer in either direction, the motor will run in the selected direction. Turning the potentiometer all the way in either direction will give us full speed in the applicable direction.

Note *In order to use a potentiometer, we need to read the position of the potentiometer wiper to get a value we can input into our control scheme. We will be using an 8-bit value for the potentiometer so the reading will go from 0 to 255. We will select 128 as the point that gives us zero power to the motor.*

The output to the motor driver will be a direction bit and a PWM value. These values are sent to the motor amplifier/driver. Exactly how this is managed in the driver is a function of the motor driver we use, but most drivers have the following three control wires for each motor:

- Direction bit
- Enable/inhibit bit enables the driver, brake bit
- PWM input for speed

We can extract the direction and PWM by interpreting the 0 to 255 value of the potentiometer as follows:

- Set the *direction* bit as follows:
  - If the value is below 128, set direction to negative. Set direction bit to 0.
  - If the value is 128 or above, set the direction to positive. Set direction bit to 1.
- Set the *speed* so it will always be 0 at 128:
  - If the value is 128, set the PWM value to 0.
  - If the value is above 128, set the PWM value to (Pot value-128).
  - If the value is below 128, set the PWM value to (128-Pot value).

We decided earlier that we will use the two-axis amplifier made by Xavien. This is a very easy to use and fairly powerful amplifier that can handle up to 6 amps maximum at 55 V maximum for short periods. It readily accepts the three signals that we need to control the motor. Having two axes on this amplifier is a useful convenience that will allow us to use this same amplifier to run our stepper motors in a later part of this book.

PWM Frequency Considerations

The frequency that we use for the PWM signal is selected so that it is above the hearing range of human beings and domestic animals. The noise is caused by loose laminations and other magnetically sensitive components in the motor. High square wave frequencies are extremely irritating to the human ear and are to be avoided. As far as the control of the motors goes, 60 Hz is completely useable. We will select 20,000 Hz or so for our frequency, though at the power we are using, 2000 Hz would also be acceptable because these little motors do not have a lot that will start vibrating in them at our power levels. However, do keep this in mind when you need to run a larger motor.

Note *Most industrial amplifiers run at 40,000 Hz to keep the noise that may be generated above the hearing range of domestic animals.*

The circuitry needed to run our motor is shown in Figure 14.2. This circuitry reflects what needs to be wired and where to run the motor with the Xavien amplifier and a 16F877A microcontroller. The circuitry follows the scheme used on the LAB-X1 so that we can use the LAB-X1 as received from the manufacturer for our experiments.

The wiring shown in Figure 14.2 follows the wiring for the LAB-X1 so that the LAB-X1 can be used as the motor controller. It is not desirable that we build a standalone controller at this point in our learning experience.

Connections to the Amplifier and Processor

On the 16F877A the continuous, background HPWM signals that we need to run our motors are available only on PORTC and only on pins C1 and C2. We will use pin C1 as the PWM pin. We will use pin 3 for direction control to keep the pins together on PORTC. Since we need to control the brake line also, we will connect it to pin C4. When done this way we will be controlling the motor exclusively from PORTC. You will need three wires, about 12 inches long, with push on connectors on each end to connect the LAB-X1 to the Xavien 2-axis amplifier.

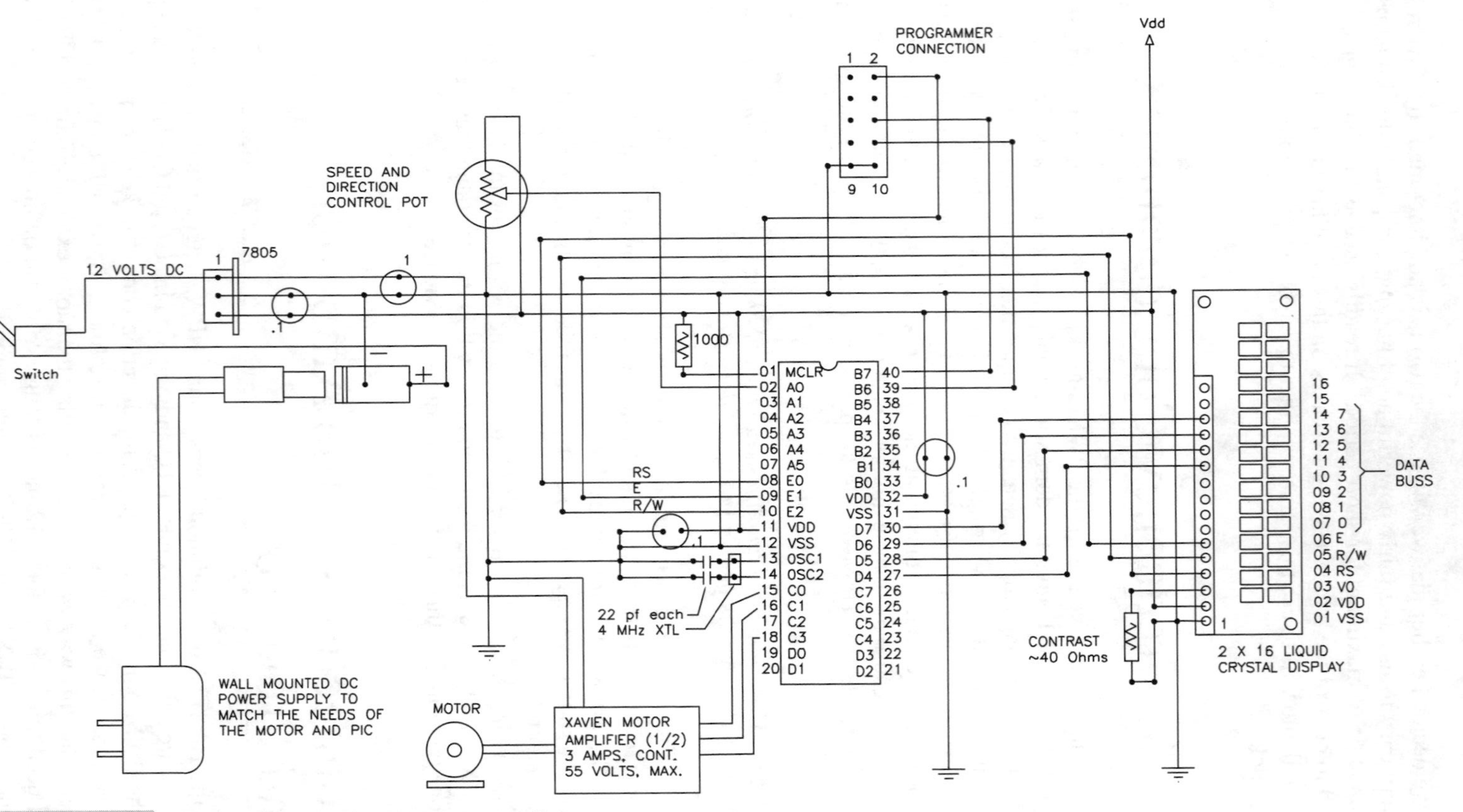

Figure 14.2 Wiring diagram for motor controller

For power, we need two wall transformers. One will provide 9-16 V at one amp for the logic supply of the controller board and the other will provide between 13 and 50 V at about one amp for the motor power supply at the amplifier. I used one providing 13.5 V at 1 amp. Each power supply should provide its positive terminal at the center and the negative voltage on the periphery of the connector. On the setup I used 2.1 mm connections were provided.

The rest is software.

The Software to Run the Motor

First, let us get the motor running. Then we will add all the other features that were discussed in the preceding paragraphs.

As we did in the first half of the book, the first thing we have to do in the software is to set up the LCD and the ports we will be using.

Let us first list the various segments of code that are needed and then we will put it all together in the proper order so all the DEFINEs are on top and so on.

As always, let us get the LCD connections defined first:

```
CLEAR                                ; clear
DEFINE OSC 4                         ; clock speed
DEFINE LCD_DREG PORTD                ; define LCD connections
DEFINE LCD_DBIT 4                    ; 4 bit protocol
DEFINE LCD_RSREG PORTE               ; Register select port
DEFINE LCD_RSBIT 0                   ; Register select bit
DEFINE LCD_EREG PORTE                ; Enable register
DEFINE LCD_EBIT 1                    ; Enable bit
ADCON1 = %00000111                   ; Set Digital bits
LOW PORTE.2                          ; Set for write only
PAUSE 500                            ; Pause to let LCD start up
LCDOUT $FE, 1, "Ready and reset"     ; clear Display
PAUSE 500                            ; Pause to see the reset
                                     ; message
LCDOUT $FE, 1                        ; clear Display again
```

Next, we write the code to read the potentiometer on PORTA.0 and define the variable to store the potentiometer value that we will read:

```
POT_VAL VAR BYTE                     ; variable defined as 8 bits
                                     ; wide
```

The DEFINES for reading the potentiometer are:

```
DEFINE ADC_BITS 8          ; Set number of bits in result as 8
DEFINE ADC_CLOCK 3         ; Set clock source (3=rc)
DEFINE ADC_SAMPLEUS 50     ; Set sampling time in µS
```

The command to read the potentiometer is:

```
ADCIN 0, POT_VAL ; Read channel 0 to POT_VAL
```

The PWM control needs are as follows for port 1 as expressed on line C.2:

```
DEFINE CCP1_REG PORTC  ; Port to be used by HPWM 1
DEFINE CCP1_BIT 2      ; Pin to be used by HPWM 1
                       ; Since no timer is defined,
                       ; Timer1 will be used
HPWM 1, X, 2500        ; the command that starts the background
                       ; PWM
```

Now we are ready to define the main loop, which contains the following pseudo code:

```
MAIN:                                    ;
   Read the pot                          ; Pseudo code
   Calculate the necessary values        ; Pseudo code
   Output to PWM command                 ; Pseudo code
GOTO MAIN                                ;
END                                      ;
```

If we put all of the precedings together in proper sequences and add in the necessary odds and ends, we get the following program. Program 14.1 controls the motor from 0 to 100 percent of full speed in one direction only at this stage.

Program 14.1 Basic motor speed control program. Simple 0 to +100% power.

```
CLEAR
DEFINE OSC 4                  ; clock
DEFINE LCD_DREG PORTD         ; define LCD connections
DEFINE LCD_DBIT 4             ;
DEFINE LCD_RSREG PORTE        ;
DEFINE LCD_RSBIT 0            ;
DEFINE LCD_EREG PORTE         ;
DEFINE LCD_EBIT 1             ;
DEFINE ADC_BITS 8             ; Set number of bits in result
DEFINE ADC_CLOCK 3            ; Set clock source (3=rc)
DEFINE ADC_SAMPLEUS 50        ; Set sampling time in µS
DEFINE CCP2_REG PORTC         ; Port to be used by HPWM 1
DEFINE CCP2_BIT 1             ; Pin to be used by HPWM 1
                              ; Since no timer is defined,
                              ; Timer1 will be used
ADCON1 = %00000111            ; Sets ports A and E to digital
LOW PORTE.2                   ; Set LCD to write mode only
TRISC = %00000000             ;
```

(continued)

Program 14.1 Basic motor speed control program. Simple 0 to +100% power. (*continued*)

```
PORTC.0 = 0                    ; turn off brake
PORTC.1 = 1                    ; PWM LINE
PORTC.3 = 1                    ; Direction of motor
POT_VAL VAR BYTE               ; Variable for the potentiometer
PAUSE 500                      ; Pause ½ sec for LCD startup
LCDOUT $FE, 1, "Ready and reset" ; clear Display
PAUSE 500                      ; Pause for message display
LCDOUT $FE, 1                  ; clear Display again
MAIN:                          ;
  ADCIN 0, POT_VAL             ; Read channel 0 to POT_VAL
  HPWM 2, POT_VAL, 20000       ; Put it in the PWM command
  LCDOUT $FE, $80, "Speed=",DEC3 POT_VAL ; Display speed
  LCDOUT $FE, $C0, "Direction=1" ; Display direction
GOTO MAIN                      ; Return to loop
END                            ; All programs must end with END
```

When you get this program running and the motor responding, you have the basic control of the motor under your control. You will notice that there is an appreciable deadband near the zero speed of the motor. Motor startup is delayed. This can be taken care of by adding an appropriate value to the equation of motion.

The rest is making the software more sophisticated to reflect the control we have in mind. Let us do just that in Program 14.2.

The first thing we need to do is to determine the power settings at which the motor starts to turn in each direction and the settings we want to use for the maximum speed in either direction. We can read all this from the LCD in Program 14.1 as we run the motor. The value at which the motor starts and the maximum speed will depend on the motor you are using. Write them down for this motor for future reference.

In our control scheme, the pot reading is to be interpreted in three ways depending on whether the reading is 128, below 128, or above 128. The direction bit is set as determined by this value.

At below 128, we limit the maximum speed by multiplying the pot value by a factor less than 1.0. Doing this will give us the full speed we want at the full travel of the pot. Also, we have to use a minimum value that will start the motor moving as soon as we go below 128. This is done by adding the minimum value to the reading. Combining these in an equation we get:

```
Power = multiplier * (127 - POT reading) + minimum power
needed to start motor.
```

At 128 we turn the motor off.

At above 128, we first subtract 128 from the pot value and then treat the value the same as we did for below 128. The equation we get is as follows:

```
Power = multiplier * (255 - pot reading) + minimum power
needed to start motor.
```

When all of this is incorporated into the control scheme, the PBP code that we get for the main loop in the program is as shown in Program 14.2 (however the multipliers have not been incorporated).

Program 14.2 Comprehensive DC motor control. (No encoder or other feedback.)

```
CLEAR                          ;
DEFINE OSC 4                   ; clock
DEFINE LCD_DREG PORTD          ; define LCD connections
DEFINE LCD_DBIT 4              ;
DEFINE LCD_RSREG PORTE         ;
DEFINE LCD_RSBIT 0             ;
DEFINE LCD_EREG PORTE          ;
DEFINE LCD_EBIT 1              ;
DEFINE ADC_BITS 8              ; Set number of bits in result
DEFINE ADC_CLOCK 3             ; Set clock source (3=rc)
DEFINE ADC_SAMPLEUS 50         ; Set sampling time in µS
DEFINE CCP2_REG PORTC          ; Port to be used by HPWM 1
DEFINE CCP2_BIT 1              ; Pin to be used by HPWM 1
ADCON1 = %00000111             ; Sets ports A and E to digital
LOW PORTE.2                    ; Set LCD to write mode only
TRISC = %00000000              ; all outputs
PORTC.0 = 0                    ; TURN off BRAKE
PORTC.1 = 1                    ; PWM LINE
PORTC.3 = 1                    ; DIRECTION OF MOTOR
                               ;
POT_VAL VAR BYTE               ; Variable for the potentiometer
MOT_PWR VAR BYTE               ;
PAUSE 500                      ; Pause 0.5 seconds for LCD startup
LCDOUT $FE, 1, "READY AND RESET"  ; clear Display
PAUSE 500                      ; Pause for message display
LCDOUT $FE, 1                  ; clear Display again
                               ;
MAIN:                          ;
   ADCIN 0, POT_VAL            ; Read channel 0 to Pot_Val
   SELECT CASE POT_VAL         ; Implement the decisions
   CASE IS <128                ;
   MOT_PWR = 127-POT_VAL       ;
   PORTC.3 = 0                 ;
   CASE 128                    ;
   MOT_PWR = 0                 ;
   CASE IS >128                ;
   MOT_PWR = POT_VAL-127       ;
   PORTC.3 = 1                 ;
   CASE ELSE                   ;
   END SELECT                  ;
```

(continued)

Program 14.2 Comprehensive DC motor control. (No encoder or other feedback.) (*continued*)

```
    HPWM 2, MOT_PWR, 20000  ; Put it in the PWM command, line C1
    LCDOUT $FE, $80, "SPEED=",DEC3 POT_VAL  ; Display speed
    LCDOUT $FE, $C0, "DIRECTION=",DEC1 PORTC.3  ; Display
                                                ; direction
GOTO MAIN                   ; Return to loop
END                         ; All programs must end with END
```

Program 14.2 provides the comprehensive control of the DC motor we are looking for. We are controlling the speed, direction, and power setting limits for the motor.

15

RUNNING DC MOTORS WITH ATTACHED INCREMENTAL ENCODERS

In this chapter we will learn how to control DC motors that have simple two phase encoders attached to them to tell us how fast the motors are moving and how far they have moved (see Figure 15.1). In order to do this and at the same time see what is going on we need fairly coarse exposed encoders that allow us to see their movements an encoder count at a time as we give the motors the instructions to move. Using these coarse encoders is extremely useful for the learning process but is not the best solution for industrial applications. As we will see this is so because the error signals we use are based on how many encoder counts we are from our target position or speed and

Figure 15.1 Small DC electric motors with encoders that show simple two phase, quadrature encoders and encoder readers attached to them.

the more encoder counts there are per revolution, the better the control we have over the process. Let us proceed with this in mind.

Changing the Processor in the LAB-X1

Let us now replace the 16F877A in the LAB-X1 with the 18F4331. Since this chip will keep track of the encoder counts automatically, it will allow us to concentrate on the other aspects of software for controlling encoded motors. Getting the encoder position automatically whenever we need it by simply reading two memory locations in the 18F4331 is a tremendous help in controlling motors with encoders at our level of expertise.

Note *In the following programs, if you run the program and the motor runs away, out of control, it means that the motor needs to have its direction reversed. Reverse the wires to the motor to fix the problem. The problem can also be fixed by reversing the two leads for the encoder signals, but it is more difficult to do that. If you prefer fixing the problem in software, the signals to PORTC.3 have to be reversed.*

The MCU controls the motor through PortA and PortC as follows:

The encoder signals will be connected to PORTA.3 and PORTA.4 (see the data sheet, page 111).The motor amplifier will be connected to PORTC as usual, as follows:

Brake	PORTC.0	Made low to turn off the brake and thus enable the amplifier. This could be tied low in hardware.
PWM	PORTC.1	
Not used	PORTC.2	This is the other PWM signal (not used).
Motor direction	PORTC.3	

Set the LAB-X1 for 20 MHz operation by moving the ABC jumpers to 2-1, 2-1, and 3-2. The 18F4331 can be run at 40 MHz, but the LAB-X1 board is limited to 20 MHz.

It would be worth your time to get the data sheet for the 18F4331 and scan the pages in Chapter 2 on setting the processor up and Chapter 16 on encoder capture. It is also worth rereading Chapter 10 on this processor if it is not fresh in your mind as you begin to experiment with encoded motors.

Confirm that the jumpers on the LAB-X1 have been moved to run it at 20 MHz. Open the meProg.exe program for the programmer and set all the variables for the programmer to the values given in Table 10.1. If these values are not right, the programmer will not be able to program the PIC 18F4331 properly. In Program 15.1, the motor gain is set to 18 to make sure that the motor will actually move back to the zero position. At this stage we are not using a proportional or integrating function to increase the gain to ensure this.

If the motor runs away there is a wiring error, it means that the motor is connected backward to the way we want it to be. Reverse the motor leads to make it move in the right direction to correct the error. (This can, of course, also be done in the software at the motor direction control bit which is PORTC.3 as mentioned earlier.)

DC Servo Motors with Encoders

When we get really serious about running motors with microprocessors, it is understood that we are talking about running motors that have optical encoders attached to them. This arrangement allows us to control the speed of the motor and its absolute position at all times. This is what is needed to realize the rapid changes in speed and position that are necessary to build sophisticated multidimensional positioning machines like pen-based plotters, laser cutters, robots, and CNC machines. For the hobby robot enthusiast, the needs of the robot are more fully met by this arrangement than can be met by any other type of motor control arrangement.

At our level of experimentation and learning, our interest is in the control of small motors that have relatively coarse encoders attached to them. These encoders provide a two phase signal, where one phase leads the other by 90 degrees in a normal 360 degree cycle. The usual signal is a square wave, as illustrated in Figure 15.2. Staggering the signals in this way allows us to determine the direction of rotation of the motor by determining which phase is leading.

A third channel can be added to provide an indexing pulse once during each revolution. The edge of this pulse can be used to position the motor exactly within a revolution of motion. Having this one repeatable starting position allows all other motor positions to be duplicated exactly. The encoders we are using do not have this third signal.

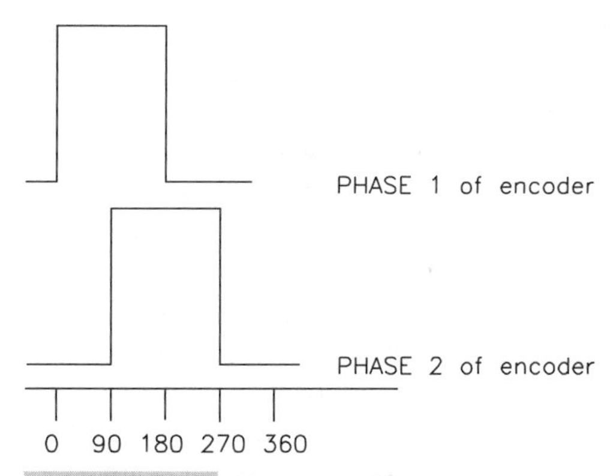

Figure 15.2 Encoder signals: one signal leads the other by 90 degrees in a 360 degree cycle.

Note *A microswitch has to be used in conjunction with the index pulse to find the revolution that represents the home position for one axis. If the index pulse and the encoder count conflict such that the microswitch switches near the full count (255 for a 256-slot encoder), an offset has to be added to the encoder count to make sure that the right indexing revolution has been identified.*

Using encoders effectively is a complicated business with many interdependent variables that need to be kept in mind when designing a control algorithm. We need to proceed one step at a time in an orderly manner if we are to understand what we are trying to do.

I have provided a number of programs to demonstrate the conditions that are encountered when encoders are in use. The programs get more complicated as we proceed. Each program allows you to play with one of two of the variables. The programs build upon each other to demonstrate the use of encoders to control the behavior of the motor. In the final programs we learn to use interrupts to ramp the motor up and down and the make a controlled move in which the ramp up, the run and the ramp down are under our control and the length of the move is defined.

WORKING PARAMETERS

Let us agree that we will always work with moves that take a few seconds and that we will ramp up for one or two seconds and ramp down for one or two seconds within this move. I have selected these parameters to make it easy for us to see the three phases of the motor's operation and still have the move completed in a reasonable time.

We will write some of our programs so that the motor runs back and forth continuously with a 1-second pause between reversals. This will allow us to vary the operational parameters and see what happens as we play with them without having to restart the program again and again.

The motor I used had 42 slots in its encoder. At 1500 rpm this motor would traverse 63,000 encoder counts in a minute. Let us agree on moves of a few thousand encoder counts for most of our experiments to keep it in round numbers.

First let us consider breaking the move into smaller segments to see what the effect might be. We break the move up into smaller segments so we can assign a speed and distance to each segment and thus achieve the move profiles that we are interested in. The simplest profile is the move that ramps up and down at a controlled rate at each beginning and end of the move. The major benefit of this scheme is that it allows us to make coordinated moves in which all the motors start, ramp up, run, and ramp down in the same time frames without regard to how long the move for each motor is.

The usual case for this is two (or more) motors where one motor makes a short move and one motor makes a long move, but they do it as a coordinated move where the motors start, ramp up, run, ramp down, and stop together. If we were to plot one move against the other, we would get a line that describes the desired profile/path. All this means is that we have to be able to ramp a motor up or down at any rate specified and run it at any rate specified. Only then will we be able to make coordinated or straight line move in any desired sequence. Following a curved path is a bit more complicated in that the moves for all the motors have to be broken up into little segments and executed simultaneously, and this takes time.

Now that we know what is required for a multimotor system, let us first concentrate on the control of just one motor, keeping in mind that multimotor moves are just a number of motors moving together with each one following its own move profile on a common time schedule. If we can make one motor follow any profile we can describe, we can control more than one motor in the same time frames and create more complicated moves.

Having made the motor move from point to point and seen the effect of the gain on the move and its stability at the move destination, we move to adding controlled ramping up and ramping down to the start and end of the move. In order to do this we have to decide on either the rate of the ramping or on the time that the ramping is to take. First let us look at some timing and gain constraints that are going to be imposed on us in any digital system before we decide how we are going to do this.

Gain

If we read a potentiometer with a resolution of 8 bits, the smallest gain that we can specify is zero and the largest gain we can specify is 255. We know that there is some friction in the system that will not allow the motor to move if the gain is a very small. Let's assume for now that this value is 14 or so for our motor. At a gain of 15, the motor moves at the lowest speed that we can make it move. (There are schemes to make it move slower, but our processor will not be fast enough and our encoder is too coarse to do that, so we will not consider that complication here.) Let us further assume that at a gain of 15, the motor moves at 20 counts per second (about half a revolution per second). What does this mean with respect to the control algorithm we are designing?

It means:

1. Obviously, we cannot send a move of less than one count at a time (though a zero move might be used as a no-op command under certain circumstances).
2. The algorithm has to be ready to send the motor the next command before it completes the one-count move. This in turn means that finer encoders, with larger counts, require the use of faster processors because one count for high count encoders goes by very, very quickly.
3. Finer encoders will allow us to run the motor slower because one count on them represents a smaller fraction of a revolution, and that is the smallest move we can command.
4. As a rule, the slower you want to run, the faster the processor you need and the higher the counts the encoder has to have.
5. The gain of the amplifier must be tied to the error (times a multiplier) between the actual position of the encoder and the commanded position of the encoder. This has an interesting corollary in that it implies that the faster we want to run the motor, the greater the positional error that has to be tolerated.

Note *If the motor is at the position it is supposed to be, the amplifier has to be off. We need to provide power to the motor only if it is not where we want it to be, and the farther we are from the destination, the larger the power provided has to be. The integrating portion of the PID loop tries to manage this, but it is not perfect. (Thus the need for the derivative part of the loop.)*

6. If the gain multiplier is too high, we can expect the system to go into an uncontrollable oscillation at the end of the move or even to overshoot during the move and cause an irregular speed profile. Controlling this depends on how often we can issue a corrective command to the controller, meaning again that faster processors have advantages we need. In any case, we need a command to be issued before the last command is completed or the motor will have stopped in midmove.
7. All in all, there are a number of compromises of which we have now become aware of. Making the best compromise between the many is the task at hand, so we will write a program or two to investigate what all this means.

Before we start discussing the control of motors that have encoders attached to them, we need to understand a few things about how the control is implemented and we need to understand a few terms that are used with encoders and servomotors.

The mysterious PID loop will be discussed in detail and explained in simple English. Each of its features will be implemented and built upon in successive programs so that you can see exactly how this is done in an 8-bit system with integer 8/16 bit math. With an encoder attached to the motor, the abilities we have as regards controlling the motor are greatly enhanced. Now, not only can we tell how far the motor has moved, we can also tell how fast it is moving, and so we can control the trajectory that it will follow. In order to implement this control we have to be able to master a number of competencies. Among them are the ability to perform the following functions. A program is designed for each example.

- **Program 1** Holds a motor at any one encoder position, no gain changes
- **Program 2** Holds a motor at any one encoder position, adds proportional gain
- **Program 3** Holds a motor at any one encoder position, adds proportional and integral gains
- **Program 4** Determines motor speed versus motor gain characteristics
- **Program 5** Determines stopping characteristics of motor when the power is turned off
- **Program 6** Controls speed and direction from potentiometer
- **Program 7** Adds and subtracts potentiometer readings to and from target position
- **Program 8** Runs motor back and forth from potentiometer reading
- **Program 9** Ramps motor up and down for a given time
- **Program 10** Moves motor back and forth a fixed number of counts
- **Program 11** Controls move with specified parameters
- **Program 12** Controls motor position with radio control signals
- **Program 13** Uses the servo exerciser program

- **Program 14** Causes motor to act as a radio controller servo
- **Program 15** Provides another way to control the speed of a motor from a radio controlled signal

DISCUSSION OF PROGRAMS DEVELOPED AND PID LOOP

There is a considerable mystique attached to the running of DC motors with optical encoders. In this chapter we will endeavor to understand what this mystique is all about by creating a number of simple, experimental programs that will illuminate the problems encountered and the solutions to them. We will begin with a simple program that simply holds a motor at any given position and returns the motor to that position if it is disturbed. We will go on to a final program in which we specify the final destination as move components to be attained and the speed at which it is to be attained; the program will do the rest. We will not cover the creation of complicated motion profiles, but you will be able to create these for yourself once you understand the basics that we will cover.

The first series of programs demonstrate the basic techniques used to run motors in the simplest way that I could think of. The later programs demonstrate how the various features available in the microprocessor are used to provide a more sophisticated approach to the problems at hand. This is done with timers and other hardware and software attributes of the microprocessor, which are explained and demonstrated in detail.

Using a coarse external encoder allows you to see the encoder move a single count. It gives you more time to implement an encoder count counting routine if you decide to implement one in software. When you are adding a motor to a real-world situation, a high count encoder with an IC that can keep track of the encoder counts in hardware is a better solution. We will not be implementing any schemes for keeping track of the encoder counts in software in this text. If you are interested in more information on this, go to the Microchip Technology web site and see the section on motor control.

Optical Encoder Information

The information that we get from the motor encoder is a function of the number of encoder slots in the encoder attached to the motor. As the encoder counts increase, we can determine how far the motor has moved more accurately and how fast it is moving in a shorter time. However, the time we have to read the encoder between states get shorter and shorter with increasing encoder counts, and this is critical. It becomes necessary to keep track of the encoder counts with hardware rather than software.

> **Note** *High count encoders are more expensive and more fragile, but they are very useful.*

There is a direct relationship between the number of encoder counts and the speed and position of the motor that is being controlled. The ideal situation is to have the smallest number of encoder counts that will do the job. This has to do with how closely the motor has to be positioned and how closely it has to follow the trajectory profile that we are interested in. We have to make changes to the profile often enough to meet the tolerance specification. The motor will tend to depart from the specified motion profile at a certain variable rate depending on the changes in the path and load conditions. The power input correction has to be made often enough to keep the motor within acceptable error for the speed and the distance moved (position). In short, having a large encoder count makes it possible to make more precise corrections more rapidly.

We are going to use the PIC 18F4331 in the LAB-X1 as the controller for our motor. This PIC can be run at 40 MHz, which is twice as fast as most PIC CPUs. This PIC has the added ability of keeping track of the encoder position in hardware without any effort on our part. The feature is a tremendous advantage because the encoder has to be read constantly so you do not lose a count. This takes up a tremendous amount of processing time and requires well-developed programming skills if done in software. It also takes a suitable, fast language to keep track of the encoder and at the same time run the motor as it executes the program. A fast processor is a must to do all these things. Though the 18F4331 can be run at 40 MHz, we are limited to 20 MHz on the LAB-X1 board.

Here is a more expansive list of the 15 programs that will be created in this chapter with a short description of what each program demonstrates or accomplishes. The development is progressive.

- **Program 15.1** A simple program to hold a motor at a position. This is a basic requirement. The motor must hold the position commanded to be usefully deployed. This is the holding program in its absolutely simplest form. The motor gain is fixed at a small value. just enough to overcome the system friction.
- **Program 15.2** A program to hold a motor at a position with proportional gain added to the return algorithm based directly on how great the positional error is. This adds sophistication to the holding loop. Now the motor will have a higher gain the farther it is from its target position. It will now reach its positional goal faster. The gain is modified each time through the move loop till the target position is reached. There should be no overt problems with this program even if there are large load changes.
- **Program 15.3** An improved program to hold a motor at a position with gain determined by a more sophisticated but simple algorithm based on a SELECT CASE loop. The gain will be limited and tunable within the SELECTIONs. Now the motor returns home without going out of control. It is important to note that the system has to be tuned to the response of the specific motor being used. This is a more sophisticated implementation with both the proportional function and the integral function. The SELECT CASE loop is being used in the place of an equation here! We need to learn how to do this for all sorts of conditions where one needs to implement the result of an equation in a program.
- **Program 15.4** A program in which a potentiometer is used to control the speed of the motor. The potentiometer value and the speed of the motor are displayed on the LCD so that we can gather the data to be plotted. This allows us to look at the gain-speed relationship as a plotted function. The results are shown in Table 15.1.

You should be able to implement simple techniques like this to gather data and to determine what is going on in your experiments on a day to day basis. We will refer to this data plotted during this experiment in our work in this book from time to time when we need to implement proportional relationships in our gain algorithms. This is also how we determined the friction factor "K" for the PID loop for the motor under consideration.

- **Program 15.5** A program to determine how many encoder counts it takes the motor to stop from any given speed. The information gained is used to design our stopping algorithms in later programs.
- **Program 15.6** A program in which motor speed and direction is controlled by the potentiometer. No ramping control is provided other than by the potentiometer.
- **Program 15.7** This program is modified in such a way that the potentiometer reading is now added and subtracted from the target position the motor is trying to achieve a zero error condition. In doing this the motor speed is controlled by the error between the target position and the actual position as determined by the potentiometer. This is rudimentary speed control.
- **Program 15.8** In this program, the motor runs back and forth for a given arbitrary distance and the gain is controlled by the potentiometer. There is no ramping at the ends of the moves. The effect of the gain on the motor motion can be examined and recorded.
- **Program 15.9** In this program, the motor ramps up for one second and ramps down for one second. Ramping is controlled by interrupts that occur every 100 microseconds. The gain is modified up or down during each interrupt. Move distance is not specified. The program demonstrates an orderly ramp down of the motor speed. There is a basic need to be able to stop at the end of a motor run in an orderly fashion, and this program demonstrates the basic techniques used. It ramps up to speed and stops again and again automatically. You can play with the ramping rate with the potentiometer to see what happens.
- **Program 15.10** In this program, the motor moves back and forth 2500 counts and ramping is implemented both on speeding up and on speeding down. No adjustments of any kind are permitted in this program. You can see the encoder counts on the LCD as the motor goes back and forth.
- **Program 15.11** A program that causes a controlled move with ramping up and ramping down with specification of the length of each move in each interrupt. This is the first program where we have complete control over the motion of the motor.
- **Program 15.12** This program uses a radio control signal to control the position of a motor as it runs back and forth. The control being implemented is the position of the motor. The control of the signal is a standard radio control hobby radio signal pulse width from 750 to 2250 microseconds.
- **Program 15.13** This is a program for the servo exerciser. Servo exerciser provided signals that are equivalent to an R/C receiver. Other signals can be simulated on this exerciser as needed for our experiments.

- **Program 15.14** This program turns the servo motor into a radio controlled servo where the motor position is proportional to the pulse width received.
- **Program 15.15** This program demonstrates another way to control the speed of a servo motor with a hobby radio control signal.

New Terms

Before we start on the programs, we need to discuss a few new concepts so we are all on the same page:

- **Encoder** An optical encoder that provides a two phase signal that we can use to determine how far the motor has moved and the direction in which it is moving. The information received also allows us to determine the speed of the motor. In our particular case, we will be using an encoder with 42 slots. All our discussions will be based on encoder counts as compared to revolutions per minute so that we do not have to undertake any conversions.
- **Servo** An electric motor that can be programmed to follow a signal. The word *servo* has the same root as the word *servant,* and in our case the motor is acting as a servant to the error signal that we introduce into the system. The error signal itself is the difference in encoder counts between where the motor is and where we want the motor to be. We control the operation of the motor by constantly adding to and subtracting from the error signal to create the motion profile that we desire.
- **Integer mathematics** The microprocessor and the language that we are using is limited to using 8- and 16-bit variables and integer mathematics. We do not have a way to solve an algebraic equation and use its results within our control algorithms. However, simple relationships can be made to serve some of our needs, and the SELECT CASE construct can be used to effectively provide the kind of relationships that an algebraic equation delivers. We will demonstrate the use of this construct to control the speed of the motor as determined by the error signal in a number of ways.

The PID Loop Explained in Simple English: The PID Control Equation and Its Components

The usual scheme used to control an encoded DC motor is called a PID loop. In the equation that represents the gain/motion of the motor, the P, I, and D represent the three basic components of the feedback loop. A constant, K, is needed to take care of the overall friction in the system. In layman's terms these variables are defined as follows:

- K (when used) is a constant needed to represent the overall system friction.
- P represents the proportional part of the control loop.
- I represents the integrating function in the control loop
- D represents the derivative part of the feedback equation.

Before we go any further, let us get an understanding what we are talking about when we say that the motor is controlled by a "PID" loop or equation. The PID loop defines how much energy is to be fed to the motor at any instant during a move. This is based on where the motor is and where it was expected to be. As just stated, there are four parts to the equation that determines this load. The three main components are referred to as the P, I, and D, and the minor friction component is referred to as K. If these four components are described properly within the control algorithm, and if a proper encoder has been selected, much improved control of the motor will be achieved.

Let us look at the components, one at a time, to see what their functions are and what they accomplish. The control scheme that we develop does not have to be mathematically perfect to give us good performance. In fact, with PBP and its limited 8- and 16-bit variables and integer math, a mathematically perfect system cannot be achieved. However, we can get close enough to have acceptable operation.

The Friction Component: K Because the motor does not start moving until it has overcome the friction in the system, a certain amount of power has to be added to the system before the motor will start to move. This is the constant K. K is often ignored because it is a minor component, and the integrating function will take care of it the first few times through the control loop. In any system with moving parts there will be some friction. In the case of a motor, even one with nothing attached and no load, there will be friction at the two shaft bearings and at the commutators brushes, and a small voltage applied to the motor will not move it. As the voltage is increased, the motor will start to move. The voltage at which the motor starts to move is the voltage needed to overcome the friction. For our purposes, it can be considered constant, although it increases as the motor speed increases. In most cases, we can ignore this increase and use a constant to represent the frictional load. Mathematically this is expressed as:

```
K = small, fixed value
```

The Proportion Component: P The component assumes that the power we supply to the motor will be proportional to the load that the motor is under. This too is not exactly accurate, but it can be defined in that way for most practical purposes. This is the largest part of the equation and so has to be picked with some care to prevent over control. The faster we want the motor to run, the larger the load and the larger the P component. In mathematical terms the energy provided can be expressed as:

```
P = load multiplied by a constant
```

If you are running a motor under a variable load, the speed that the motor attains will be approximately proportional to the load that is on the motor. Keep in mind that in our system the gain can vary from 0 to 255. We have to select the gain so that it will stay well within these limits under all conditions. We will use a suitable multiplier and then a conditional test to ensure this.

The Integrating Component: I If there are no load changes and the system response is linear (meaning that twice the speed requires twice the power), the proportional component is all we need to run the motor. However, if the system is not linear or if the load is changing, we have to add to or subtract from the gain to keep the motor at a constant speed. We have to do this a little bit at a time each time through the control loop till the motor gets to the desired speed. This is the I, or integrating component, in the equation. Because it is needed only when there is an error in the motor position, it has to be a function of this error. The higher the positional error, the more we have to add to or subtract from the power setting to make the motor speed up or slow down to where we need it to be. As mentioned previously, this is done every time we go through the control loop. In mathematical terms the energy provided can be expressed as:

```
  I = (Commanded position - actual position ) multiplied by a
suitable factor
```

The Derivative Component: D This component is a measure of the difference between where the motor is and where we expected the motor to be at any one time. This value is calculated each time through the control loop. If there is a large difference between the two numbers, we cannot wait to integrate the power in little increments but need to make a larger adjustment right away to get the motor within acceptable parameters as rapidly as possible. Exactly how much power has to be added is a function of the system inertia, the load, and how tightly the trajectory specified for the motor has to be followed. In a metal cutting CNC machine, the tool has to follow the specified path very closely so these corrections must be made very frequently. The equation is best designed as a time-based function where we have an equation that tells us where the motor is supposed to be at any one time in each move. We can then read the actual position, compare it to what the equations tells us and make the correction. In our case the 8/16-bit math and 20 MHz processor do not lend themselves to the task at hand with ease. Even so adequate approximations can be implemented and a working system achieved.

To determine D, we need to know where the motor should be and where it actually is. The difference is the error. We want this error to be as small as possible, and our response is based on how small this error should be. If we are running a very accurate positional system we may need to look at this many hundred times a second and make constant adjustments to the load. A high count encoder is desirable when rapid adjustments have to be made. The high counts allow us to get a "change in position" reading more often.

```
   D = (Expected position - real position) * (constant or
variable for some kind)
```

One thing this means in simple terms is that there is no need for a change in the power input if the motor is where it needs to be and is moving at the desired speed.

Simulating an Equation with the SELECT CASE Construct Suppose we need to know the square of all the numbers between 0 and 4 in our control scheme, and our operating system does not support mathematical functions. We can solve the problem with the SELECT CASE statement. Each case of the number between 0 and 4 has a corresponding value. These values can be put in a SELECT CASE construct as follows to solve the equation Y=x^2:

```
SELECT CASE X
  CASE 0
    Y=0
  CASE 1
    Y=1
  CASE 2
    Y=4
  CASE 3
    Y=9
  CASE 4
    Y=16
  CASE ELSE
END SELECT
```

Since the PBP system uses 8/16 math without the implementation of the minus sign or the decimal point, we have to work around these handicaps also.

First, let's work around the minus sign. Suppose we are trying to get a motor to a designated position and our pseudo code for doing so is as follows:

```
If it is not there yet we have to keep going.
If it is at the position we have to stop.
If it goes past that position we have to reverse it.
```

If we are to make a decision on the basis of the motor position, we have to implement the decision process as follows because we cannot use a negative value. In integer math, 128–129 is not –1, it is 255. This forces us to use comparisons between the values. If we need the difference, we have to first determine which value is larger, then determine the difference and then the sign. The sign in this case gives us the motor direction bit. In the following code, we determine whether or not we are going to run the motor and, if we are going to run it, the direction in which we are going to run it.

```
TARGET = 128
POSITION =read from position register
SELECT CASE POSITION
  CASE IS < TARGET
    MOTOR DIRECTION = 1
    TURN MOTOR BRAKE OFF
  CASE IS = TARGET
    TURN MOTOR BRAKE ON
  CASE IS > TARGET
    MOTOR DIRECTION = 0
    TURN MOTOR BRAKE OFF
END SELECT
```

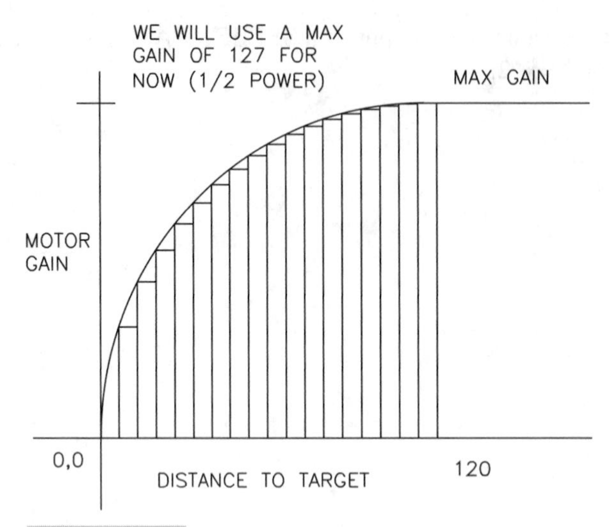

Figure 15.3 **Distance to target versus gain**

Then we have to determine how much power we should give the motor based on how far the motor is from where it needs to be. The gain will be a function of the positional error.

With the information in Figure 15.3 in mind, we can implement the equation to find the gains we need in the following SELECT CASE construct. Not every column has to be implemented to get a useable approximation of the equation.

```
SELECT CASE DISTANCE    ;
  CASE IS >120          ;
    GAIN = 127          ; We usually use only half the power
                        ; and save the rest
  CASE IS >100          ; for use when more acceleration is
                        ; needed
    GAIN = 60           ;
  CASE IS >50           ;
    GAIN = 42           ;
  CASE IS >20           ;
    GAIN = 20           ;
  CASE IS >10           ;
    GAIN = 15           ;
  CASE IS >5            ;
    GAIN = 10           ;
END SELECT              ;
```

The values that we have selected can be fine-tuned by trial and error. We can use this and similar techniques whenever we need to implement an equation within a control algorithm.

The Programs

Let us proceed with controlling an encoded motor a step at a time as per the listing of the programs earlier in this chapter.

HOLDING A MOTOR POSITION

The first form of control that we have to establish over the motor is to be able to position the motor at a specific encoder position (or count) and hold that position under all conditions.

If the motor goes over by one count in either direction we move it back to its initial position by providing either forward or reversed current to the motor to bring it back to its initial position. In order to do this we have to be able to count the encoder signals as they are generated. The processor can tell which direction the motor is moving by determining which of the two phases is leading. Once we have established the control we want, the motor will hold to this one position and return to this position if disturbed. The gain should be selected such as not to overshoot the holding position.

How the motor returns to its set position is determined by the sophistication of the control algorithm, the goal being a smooth and rapid return to the zero position without overshoot. (This is called *perfect dampening*.) If the gain is too high, meaning that too much power is applied, the motor will start to oscillate wildly. If not enough power is applied, it will not get to the zero position rapidly enough or it will not get there at all.

In a sophisticated control scheme, the power to return to zero should be a function of how far from the zero position the motor is. It should be changed in real time to reflect the load on the motor and any number of other factors that might be changing. What can be done is by and large a function of the speed of the processor we are using and the language we have selected to implement our control. A faster processor and assembly language can be used to do a lot more than the processor we are using with the PBP language. Even so, it will be possible to demonstrate all the techniques that you need to be conversant with.

When you design algorithms in assembly language they will be much like what we will undertake here. A lot of software is often first written in a higher level language, and once proper operation is confirmed the software is translated into either assembly code, C, or some other language that has been optimized for the task at hand.

The starting and stopping of the motors is trivial, so we will not consider it separately but rather incorporate it into our programs as we need to.

Program 15.1 in this chapter demonstrates how we use a simple algorithm to hold a motor at any given position. If we disturb the motor, the motor returns to its set position automatically. In this program we are using a fixed gain that is determined by the setting of the potentiometer. Starting out with a reading of zero on the potentiometer, we find that extremely low gains do not allow the motor to return to its set position. As we increase the gain on the motor we see the motor start to respond. As the gain gets higher, the motor starts to first jitter and then go into wild oscillations. Our goal is to have the motor returned to its original position as rapidly and as smoothly as possible. In order to do this, we need to adjust the gain of the motor so that it depends on the amount of the error in position. The pseudo code for positioning and holding the motor at any one encoder position is as follows:

```
Set an appropriate initial amplifier gain
Set the current encoder count registers to zero
Set the current position as the zero position
Read the encoder position
If it has increased, reverse the motor
If there is no change, do nothing
If it has decreased, move the motor forward
Go back and read the encoder again.
```

Refinements consist of adjusting the power to the motor based on how far it is from the desired position in real time and how rapidly it is responding to the corrections made. When the motor gets close to the desired position, the power supplied is adjusted to be barely enough to reach the motor's home position. The inertia of the motor and the load on it also play a role in determining the power supplied, meaning that the operation of the system is tuned to its most frequent load. A tuning/optimization algorithm can be built into the feedback loop being used if the processor is fast enough to have time to do this.

In Program 15.1 we will look at the lower 4 bits of PORTC on line 1 of the LCD so we can see what is happening with the motor direction bit. We will also look at the position counter on line 2 so we can see what is happening with the encoder counts as we play with the system. Place a line on the encoder wheel with an indelible pen so you can see the movement to every encoder position as it moves back and forth.

Program 15.1 Hold position, no proportional gain, no integration. Rudimentary "holding a motor on position" program.

```
CLEAR                       ; clear memory
DEFINE OSC 20               ; 20 MHz clock (40 not avail on the
                            ; LAB-X1)
DEFINE LCD_DREG PORTD       ; define LCD connections
DEFINE LCD_DBIT 4           ; 4 data bits
DEFINE LCD_BITS 4           ; data starts on bit 4
DEFINE LCD_RSREG PORTE      ; select register
DEFINE LCD_RSBIT 0          ; select bit
DEFINE LCD_EREG PORTE       ; enable register
DEFINE LCD_EBIT 1           ; select bit
LOW PORTE.2                 ; set bit low for writing to the LCD
DEFINE LCD_LINES 2          ; lines in display
DEFINE LCD_COMMANDUS 2000       ; delay in µs
```

(continued)

Program 15.1 Hold position, no proportional gain, no integration. Rudimentary "holding a motor on position" program. (*continued*)

```
DEFINE LCD_DATAUS 50    ; delay in µs
DEFINE ADC_BITS 8       ; set number of bits in result
DEFINE ADC_CLOCK 3      ; set clock source (3=rc)
DEFINE ADC_SAMPLEUS 50 ; set sampling time in µs
DEFINE CCP2_REG PORTC  ; hpwm 2 pin port
DEFINE CCP2_BIT 1       ; hpwm 2 pin bit 1
CCP1CON = %00111111     ; set status register
TRISA = %00011111       ; set status register
LATA = %00000000        ; set status register
TRISB = %00000000       ; set status register
LATB = %00000000        ; set status register
TRISC = %00000000       ; set status register
TRISD = %00000000       ; set status register
ANSEL0 = %00000001      ; page 251 of data sheet, status
                        ; register
ANSEL1 = %00000000      ; page 250 of data sheet, status
                        ; register
QEICON = %10001000      ; page 173 counter set up, status
                        ; register
INTCON = %00000000      ; set status register
INTCON2.7 = 0           ; set status register
                        ;
POSITION VAR WORD       ; set variables
MOTPWR VAR BYTE         ; set variables
                        ;
PORTC.0 = 0             ; brake off, motor control. 1=brake ON
PORTC.1 = 0             ; PWM bit for Channel 2 of HPWM
PORTC.3 = 1             ; dir bit for motor control
PAUSE 500               ; LCD start up pause
LCDOUT $FE, $01, "START UP CLEAR"   ; clear message
PAUSE 100               ; pause to see message
POSCNTH = 127           ; set counter for encoder, H bit
POSCNTL = 0             ; set counter for encoder, L bit
                        ;
LOOP:                   ; main loop
POSITION = 256*POSCNTH + POSCNTL    ; read position registers
SELECT CASE POSITION    ; Select loop
  CASE IS = 32512       ; if at position then
    MOTPWR = 0          ; turn off motor gain
    PORTD.3 = 0         ; turn off power LED
  CASE IS < 32512       ; if under shoot
    PORTC.3 = 0         ; set direction forward
    PORTD.2 = 1         ; turn on direction led
    PORTD.3 = 1         ; turn on power LED
    MOTPWR = 18         ; set motor gain
```

(*continued*)

Program 15.1 Hold position, no proportional gain, no integration. Rudimentary "holding a motor on position" program. (*continued*)

```
    CASE IS > 32512          ; if over shoot
      PORTC.3 = 1            ; set direction in reverse
      PORTD.2 = 0            ; turn off LED for reverse direction
      PORTD.3 = 1            ; turn on power LED
      MOTPWR = 18            ; set motor gain
    CASE ELSE                ;
  END SELECT                 ; end decision
  HPWM 2, MOTPWR, 20000      ; C.1 PWM signal actuation
                             ;
  LCDOUT $FE, $80, "PORTC=",BIN4 PORTC," GAIN=",DEC3 MOTPWR
                             ; display
  LCDOUT $FE, $C0, "POSITION =",DEC5 POSITION      ; display
                             ;
  GOTO LOOP                  ; go back to loop
  END                        ; all programs must end with END
```

In Program 15.1, the LCD displays the four lower bits in PORTC, the motor gain, and the position of the motor. The program holds the motor on position as soon as it comes on. If you move the motor to off position by turning the motor shaft manually, it will move back to the set position as soon as you let go. Since a positional error is *not* reflected in the gain of the amplifier (the proportional gain) in this program, the motor *does not* turn harder as you get further away from the zero position, and the motor may not return to the absolute zero position if the value of the gain selected is set too low. (If there is no integration function in the gain, the motor will not be able to return to the zero position at all for very small gains.) This has been compensated for by using a larger fixed value to the gain (MOTPWR=18) to make sure that the gain will always be able to move the motor. The value used has to be slightly more than needed to compensate for the friction component of the PID loop, as was discussed earlier.

The set up that I used is illustrated in Figure 15.4. This set up was used for all the experiments in the book including the stepper motors. I used quick connects between various components to allow me to switch from setup to setup. Everything is mounted on a piece of quarter inch plywood. A close up of the motor and its encoder is shown in Figure 15.5.

Program 15.2 will let us modify the control algorithm in Program 15.1 to make the gain dependent on how far the motor is from its home position. This will make the motor move to its target position more strongly as you turn the motor farther away from the target home. If the gain is unrestrained, it will cause the motor to over control and go into oscillations when you let go. You can watch the motor position and the gain on the LCD as you move the motor shaft back and forth. Note that the gain has to be limited to 255 to keep the gain byte from overflowing and may have to be limited at an even lower value to prevent over control oscillations.

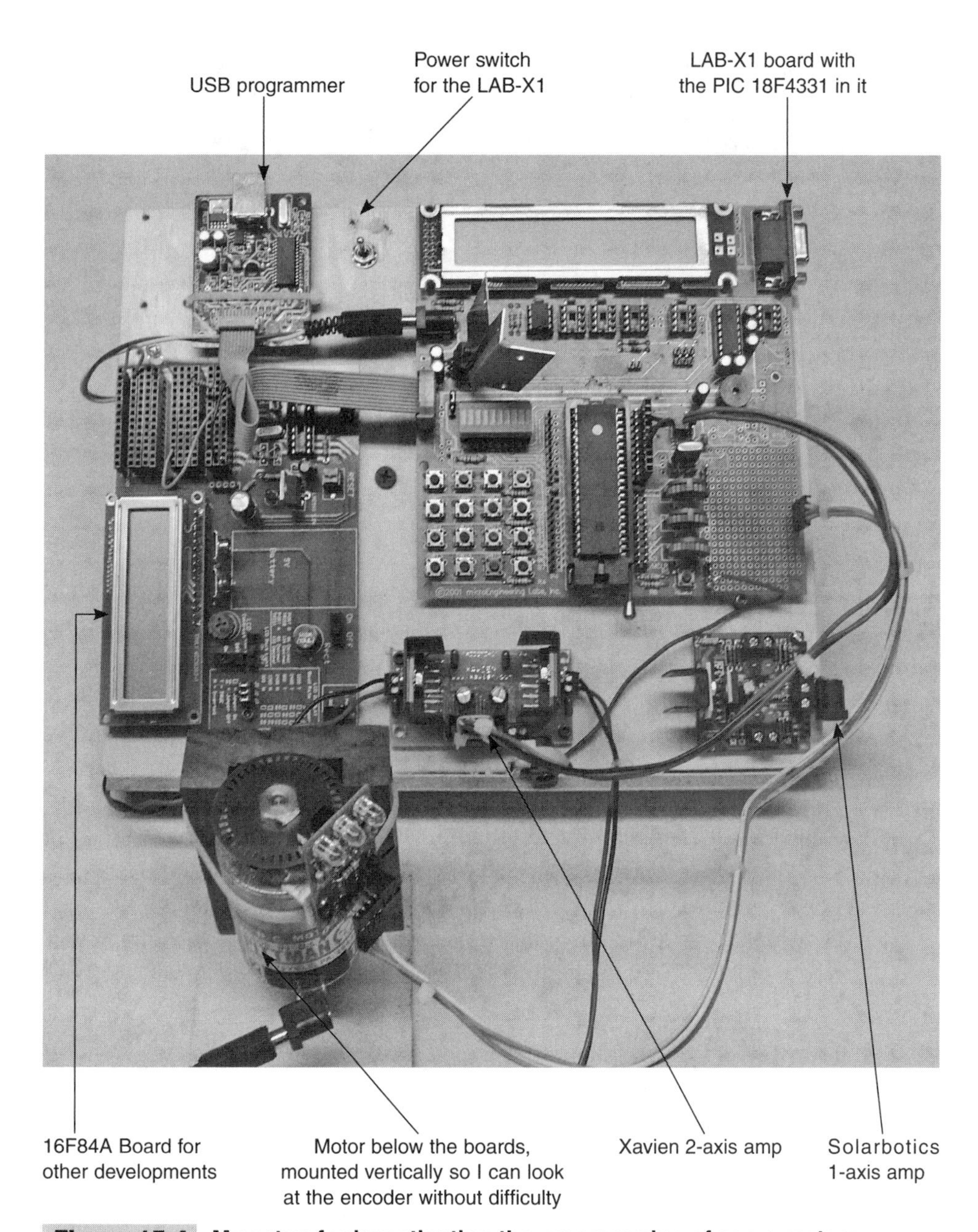

Figure 15.4 **My setup for investigating the programming of servo motors**

Figure 15.5 Detail of the motor and encoder used

Program 15.2 **Hold position, proportional gain, no integration. Improved "holding a motor on position" program.** The potentiometer controls motor position.

```
CLEAR                        ; clear memory
DEFINE OSC 20                ; 20 MHz clock (40 better)
DEFINE LCD_DREG PORTD        ; define LCD connections
DEFINE LCD_DBIT 4            ; 4 data bits
DEFINE LCD_BITS 4            ; data starts on bit 4
DEFINE LCD_RSREG PORTE       ; select register
DEFINE LCD_RSBIT 0           ; select bit
DEFINE LCD_EREG PORTE        ; enable register
DEFINE LCD_EBIT 1            ; select bit
LOW PORTE.2                  ; set bit low for writing to the LCD
DEFINE LCD_LINES 2           ; lines in display
DEFINE LCD_COMMANDUS 2000       ; delay in µs
DEFINE LCD_DATAUS 50         ; delay in µs
DEFINE ADC_BITS 8            ; set number of bits in result
DEFINE ADC_CLOCK 3           ; set clock source (3=rc)
```

(continued)

Program 15.2 Hold position, proportional gain, no integration. Improved "holding a motor on position" program. The potentiometer controls motor position. (*continued*)

```
DEFINE ADC_SAMPLEUS 50     ; set sampling time in µs
DEFINE CCP2_REG PORTC      ; hpwm 2 pin port
DEFINE CCP2_BIT 1          ; hpwm 2 pin bit 1
CCP1CON=%00111111          ; set status register
TRISA =%00011111           ; set status register
LATA =%00000000            ; set status register
TRISB =%00000000           ; set status register
LATB =%00000000            ; set status register
TRISC =%00000000           ; set status register
TRISD =%00000000           ; set status register
ANSEL0=%00000001           ; page 251 of data sheet, status
                           ; register
ANSEL1=%00000000           ; page 250 of data sheet, status
                           ; register
QEICON=%10001000           ; page 173 counter set up, status
                           ; register
INTCON=%00000000           ; set status register
INTCON2.7=0                ; set status register
                           ;
POSITION VAR WORD          ; set variables
MOTPWR VAR WORD            ; set variables
POTVALUE VAR BYTE          ; set variables
ERROR VAR WORD             ;
PORTC.0=0                  ; break off, motor control
PORTC.1=0                  ; PWM bit for Channel 2 of HPWM
PORTC.2=0                  ; PWM bit for Channel 1 of HPWM
PORTC.3=1                  ; dir bit for motor control
PAUSE 500 ; LCD START UP PAUSE      ;
LCDOUT $FE, $01, "START/CLEAR"      ; clear message
PAUSE 100                  ; pause to see message
POSCNTH=127                ; set counter for encoder, H bit
POSCNTL=0                  ; set counter for encoder, L bit
                           ;
LOOP:                      ; main loop
ADCIN 0, POTVALUE          ;
POSITION=256*POSCNTH + POSCNTL      ; read position registers
POSITION=POSITION+POTVALUE          ;
SELECT CASE POSITION       ;
  CASE IS=32513            ;
    HPWM 2, 0, 20000       ; C.1 PWM signal
    ERROR=0                ;
  CASE IS<32513            ; set motor direction
    PORTC.3=1              ; set direction
    ERROR= 32513-POSITION ;
```

(*continued*)

Program 15.2 Hold position, proportional gain, no integration. Improved "holding a motor on position" program. The potentiometer controls motor position. (*continued*)

```
   CASE IS>32513          ;
     PORTC.3=0            ; set direction in reverse
     ERROR= POSITION-32513 ;
   CASE ELSE              ;
 END SELECT               ; end decision
 MOTPWR=ERROR +14         ;
 IF ERROR=0 THEN MOTPWR=0 ;
 IF MOTPWR>100 THEN MOTPWR=100 ;
 HPWM 2, MOTPWR, 20000    ; C.1 PWM signal
 LCDOUT $FE, $80, "PRTC=",BIN4 PORTC," GAIN=",DEC3 MOTPWR
                          ; display
 LCDOUT $FE, $C0, "POS =",DEC5 POSITION    ; display
 GOTO LOOP                ; go back to loop
 END                      ; all programs must end with END
```

Playing with the setup in Program 15.2 reveals that the gain increases with the distance the motor is from its set position. Try modifying this program by adding a multiplier to the gain function. This makes the gain steeper and makes the motor more prone to oscillations. We still have the problem of the motor not reaching its home position if the error is small and the load or the friction in the system is high. Next, we will add an integration function to the motor gain to overcome that. Program 15.3 demonstrates how this is done.

TURNING POTENTIOMETER TO CONTROL MOTOR

Program 15.3 also demonstrates one of the ways in which a potentiometer (in this case acting as an error signal) can control the position of an encoded motor. In our case the potentiometer provides a count of up to 255 so we can move the motor about three turns (as set up with the 42-slot encoder, each slot in the encoder gives us two counts because the PIC 18F4331 has been set to read both the rising and falling edge of the signal).

If we want the motor to move more than three turns, we can use a multiplier to change the value read from the potentiometer, but we are still limited to the 255 different positions that the potentiometer provides. (We can also read the potentiometer as a 10-bit variable, which will give a reading of from 0 to 1023.)

As we add or subtract the potentiometer reading from the register that contains the encoder counts to bring the motor to its set position, we have a rudimentary motor control algorithm. Of course there are many improvements to be made, but basically that is what we are trying to do with all programs that control encoder coupled motors.

Program 15.3 provides the rudimentary control needed to hold a motor on an encoder position as just described. Later programs modify this code to add the features and refinements we have been discussing. As you turn the potentiometer further and further, the motor moves further and further—that is, it follows the potentiometer. The motor moves because the potentiometer position is being added to the target register (not the position register) each time through the loop. It is not cumulative. The LCD display shows the contents of the four low bits of PORTC and the 16-bit position register and the gain so we can see what is going on.

What we have is a simple motor position controller operated from a potentiometer. This would be an easy way to move a controlled device in a laboratory set up. With a few wire extensions you could also do the following:

- Move a control lever remotely
- Open and close a motorized gate
- Remove yourself from a dangerous environment
- Turn a knob or a steering wheel on a car with a large servo
- Control one axis of the orientation of a remote camera
- Use the remote motor as a general purpose positioning servo

The controlling input does not have to be the potentiometer; any of the methods that we have at our disposal to read resistances, capacitances, frequencies, and so on with the 16F877A can be used as the input signal. We can use a hobby R/C signal to create a remotely operated radio controlled system.

Let us modify the control algorithm to add one to the gain during each pass through the zeroing loop (integrating the gain) if the motor is not at its home position. This will make the motor move to its target position more and more strongly each time through the loop if is not at the target position. If unrestrained, the gain will again cause the motor to over control and go into oscillations. You can watch the position and the gain on the LCD as you move the motor shaft back and forth. Note that, as always, the gain has to be limited to 255 to keep the gain byte from overflowing. It may also have to be limited to a less aggressive value to prevent over control.

Program 15.3 Hold position, proportional gain with integration added. Sophisticated "holding a motor on position" program. Potentiometer value is added to the motor target position.

```
CLEAR                          ; clear memory
DEFINE OSC 20                  ; 20 MHz clock (40 better)
DEFINE LCD_DREG PORTD          ; define LCD connections
DEFINE LCD_DBIT 4              ; 4 data bits
DEFINE LCD_BITS 4              ; data starts on bit 4
DEFINE LCD_RSREG PORTE         ; select register
DEFINE LCD_RSBIT 0             ; select bit
DEFINE LCD_EREG PORTE          ; enable register
DEFINE LCD_EBIT 1              ; select bit
LOW PORTE.2                    ; set bit low for writing to the LCD
```

(continued)

Program 15.3 Hold position, proportional gain with integration added. Sophisticated "holding a motor on position" program. Potentiometer value is added to the motor target position.
(*continued*)

```
DEFINE LCD_LINES 2              ; lines in display
DEFINE LCD_COMMANDUS 2000       ; delay in µs
DEFINE LCD_DATAUS 50            ; delay in µs
DEFINE ADC_BITS 8               ; set number of bits in result
DEFINE ADC_CLOCK 3              ; set clock source (3=rc)
DEFINE ADC_SAMPLEUS 50          ; set sampling time in µs
DEFINE CCP2_REG PORTC           ; hpwm 2 pin port
DEFINE CCP2_BIT 1               ; hpwm 2 pin bit 1
CCP1CON=%00111111               ; set status register
TRISA =%00011111                ; set status register
LATA =%00000000                 ; set status register
TRISB =%00000000                ; set status register
LATB =%00000000                 ; set status register
TRISC =%00000000                ; set status register
TRISD =%00000000                ; set status register
ANSEL0=%00000001                ; page 251 of data sheet, status
                                ; register
ANSEL1=%00000000                ; page 250 of data sheet, status
                                ; register
QEICON=%10001000                ; page 173 counter set up, status
                                ; register
INTCON=%00000000                ; set status register
INTCON2.7=0                     ; set status register
                                ;
POSITION VAR WORD               ; set variables
TARGET VAR WORD                 ; set variables
MOTPWR VAR WORD                 ; set variables
POTVALUE VAR BYTE               ; set variables
INTPWR VAR BYTE                 ; set variables
PORTC.0=0                       ; break off, motor control
PORTC.1=0                       ; PWM bit for Channel 2 of HPWM
PORTC.2=0                       ; PWM bit for Channel 1 of HPWM
PORTC.3=1                       ; dir bit for motor control
PAUSE 500                       ; LCD start up pause
LCDOUT $FE, $01, "START/CLEAR"      ; clear message
PAUSE 100                       ; pause to see message
POSCNTH=125                     ; set counter for encoder, H bit,
                                ; 32000
POSCNTL=0                       ; set counter for encoder, L bit
                                ;
LOOP:                           ; main loop
ADCIN 0, POTVALUE               ;
POSITION=256*POSCNTH + POSCNTL      ; read position registers
TARGET =32000 +POTVALUE         ; add pot value to position
```

(*continued*)

Program 15.3 Hold position, proportional gain with integration added. Sophisticated "holding a motor on position" program. Potentiometer value is added to the motor target position.
(*continued*)

```
SELECT CASE TARGET           ;
  CASE IS= POSITION          ; at 32000
    MOTPWR=0                 ; turn off motor
    INTPWR=0                 ; zero integral gain
    HPWM 2, 0, 20000         ; C.1 PWM signal, stop motor
  CASE IS< POSITION          ; under count
    PORTC.3=0                ; set direction fwd
    MOTPWR=POSITION- TARGET +10      ; set motor gain
  CASE IS> POSITION          ; over count
    PORTC.3=1                ; set direction reverse
    MOTPWR= TARGET -POSITION +10     ; set motor gain
  CASE ELSE                  ; empty
END SELECT                   ; end decision
IF POSITION<>TARGET THEN INTPWR=INTPWR +1 ; add 1 to
                                          ; integral gain
IF INTPWR>90 THEN INTPWR=90   ; limit integral fn to 90
MOTPWR=MOTPWR +INTPWR        ; add integral to motor gain
IF MOTPWR>255 THEN MOTPWR=255 ; allow half of full power
HPWM 2, MOTPWR, 20000        ; C.1 PWM signal
LCDOUT $FE, $80, "PRTC=",BIN4 PORTC,"  GAIN=",DEC3 MOTPWR
                             ; display
LCDOUT $FE, $C0, "POS =",DEC5 POSITION+POTVALUE ; display
GOTO LOOP                    ; go back to loop
END                          ; all programs must end with END
```

In Program 15.3 it is hard to keep the motor from getting to its target position as controlled by the potentiometer. The potentiometer provides a range of motion of about three revolutions. The motor position control algorithm is essentially the basic control algorithm used in all control schemes. Major sophistication may be added, but the basic plan will remain the same. Notice that we did not implement the derivative functions in this program. (In Program 15.3 the integration value is limited to 90 to keep you from cutting yourself on the encoder wheel.)

DETERMINING THE MOTOR CHARACTERISTICS

We need to have a feel for how a motor responds to the gain that it is experiencing. In our case the gain can vary from zero to 255, and at each gain the motor will run at a certain speed. Program 15.4 allows us to use the potentiometer to input a gain from zero to 255 and to read the speed of the motor at each gain so that we can make a table of the motor response. We count the number of encoder counts seen in 100 ms. Program 15.4 displays the gain on line 1 of the LCD and the speed of the motor on line 2. A listing of the results I obtained with my motor is shown in Table 15.1.

Program 15.4 Motor gain versus speed. This program was used for determining the response data shown in Table 15.1

```
CLEAR                       ; clear variables
DEFINE OSC 20               ; 20 MHz clock
DEFINE LCD_DREG PORTD       ; define lcd connections
DEFINE LCD_DBIT 4           ; 4 data bits
DEFINE LCD_BITS 4           ; data starts on bit 4
DEFINE LCD_RSREG PORTE      ; select register
DEFINE LCD_RSBIT 0          ; select bit
DEFINE LCD_EREG PORTE       ; enable register
DEFINE LCD_EBIT 1           ; select bit
LOW PORTE.2                 ; set bit low for writing to the LCD
DEFINE LCD_LINES 2          ; lines in display
DEFINE LCD_COMMANDUS 2000       ; delay in µs
DEFINE LCD_DATAUS 50        ; delay in µs
DEFINE ADC_BITS 8           ; set number of bits in result
DEFINE ADC_CLOCK 3          ; set clock source (3=rc)
DEFINE ADC_SAMPLEUS 50      ; set sampling time in µs
DEFINE CCP2_REG PORTC       ; hpwm 2 pin port
DEFINE CCP2_BIT 1           ; hpwm 2 pin bit 1
CCP1CON = %00111111         ; set status register
TRISA = %00011111           ; set status register
LATA = %00000000            ; set status register
TRISB = %00000000           ; set status register
LATB = %00000000            ; set status register
TRISC = %00000000           ; set status register
TRISD = %00000000           ; set status register
ANSEL0 = %00000001          ; page 251 of data sheet, status
                            ; register
ANSEL1 = %00000000          ; page 250 of data sheet, status
                            ; register
QEICON = %10001000          ; page 173 counter set up, status
                            ; register
INTCON = %00000000          ; set status register
INTCON2.7 = 0               ; set status register
                            ;
MOTPWR VAR WORD             ; set variables
COUNTER VAR BYTE
SPEED VAR WORD              ; set variable
TOTAL VAR WORD              ; set variable
POT_POS VAR BYTE            ; potentiometer position
PORTC.0 = 0                 ; brake off, motor control
PORTC.1 = 0                 ; PWM bit for Channel 2 of HPWM
PORTC.3 = 1                 ; dir bit for motor control
PAUSE 500                   ; LCD start up pause
LCDOUT $FE, $01, "START/CLEAR"      ; clear message
PAUSE 100                   ; pause to see message
```

(continued)

Program 15.4 Motor gain versus speed. This program was used for determining the response data shown in Table 15.1 (*continued*)

```
LCDOUT $FE, $01          ; clear display
                         ;
LOOP:                    ; main loop
ADCIN 0, POT_POS         ; read incremental speed value
COUNT PORTA.3, 100, SPEED     ; read counter
COUNTER = COUNTER + 1    ; update counter
TOTAL = TOTAL + SPEED    ; totalize for average taken later
MOTPWR = POT_POS         ; set motor power to pot value
LCDOUT $FE, $80, "GAIN =",DEC3 POT_POS     ; display
HPWM 2, MOTPWR, 20000    ; C.1 PWM signal, channel 2
IF COUNTER = 10 THEN     ; ready to take average
  SPEED=TOTAL / 10       ; take average
  LCDOUT $FE, $C0, "SPEED=",DEC5 SPEED     ; display
  COUNTER = 0            ; reset counter
  TOTAL = 0              ; reset total
  ELSE                   ;
ENDIF                    ;
GOTO LOOP                ; go back to loop
                         ;
END                      ; All programs must end with END
```

Playing with Program 15.4 reveals the lowest gain, the value K, that is needed to get the motor turning. In my case this was 12. The program demonstrates that the speed of a motor is directly related to the gain. Figure 15.6 shows an approximate plot of the readings for my motor. The points of interest in the diagram are the friction offset the maximum speed (for half of the full gain) and the linearity of the data.

In Table 15.1, all the speeds are in encoder counts per 100 milliseconds, and we know that the motor had an encoder with 42 slots in it. Working directly in encoder counts allows us to ignore any effects that the integer math has on our results.

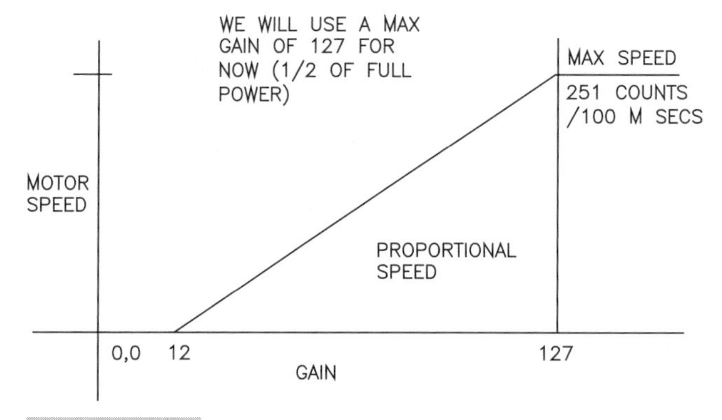

Figure 15.6 Gain versus speed: the data confirms that the response is indeed proportional; the motor speed was determined by experimentation using Program 15.4.

TABLE 15.1 ACTUAL OBSERVED VALUES

GAIN	SPEED	GAIN	SPEED	GAIN	SPEED	GAIN	SPEED	GAIN	SPEED
0	0	32	49	64	113	96	182	128	
1	0	33	51	65	114	97	183	129	
2	0	34	52	66	116	98	187	130	
3	0	35	54	67	120	99	189	131	259
4	0	36	55	68	121	100	191	132	
5	0	37	57	69	123	101	193	133	
6	0	38	59	70	126	102	195	134	
7	0	39	62	71	130	103	198	135	268
8	0	40	65	72	132	104	201	136	
9	0	41	66	73	133	105	203	137	
10	0	42	71	74	136	106	205	138	
11	4	43	73	75	139	107	207	139	274
12	7	44	73	76	141	108	210	140	
13	10	45	75	77	144	109	212	141	
14	13	46	78	78	146	110	214	142	
15	16	47	80	79	148	111	216	143	284
16	16	48	83	80	150	112	219	144	
17	18	49	85	81	152	113	221	145	
18	20	50	87	82	154	114	223	146	
19	22	51	88	83	156	115	224	147	292
20	23	52	91	84	158	116	226	148	
21	26	53	93	85	160	117	228	149	
22	27	54	94	86	162	118	230	150	
23	30	55	96	87	163	119	323	151	300
24	31	56	98	88	166	120	235	152	
25	34	57	100	89	168	121	237	153	
26	36	58	102	90	170	122	239	154	
27	38	59	103	91	172	123	241	155	307
28	40	60	105	92	175	124	244	156	
29	44	61	107	93	175	125	247	157	
30	45	62	110	94	178	126	249	158	
31	46	63	111	95	180	127	251	159	315

As a part of understanding the characteristics of the motor that we are using, we need to know how long it takes the motor to stop if the power is turned off suddenly. The information we are interested in will be expressed as a number of encoder counts. The program I wrote turned the motor on at full speed, turned it off, zeroed the position counters, and waited till the motor came to rest. It then displayed how many counts it had taken the motor to stop. This information is of interest to us in the design of the algorithm that will ramp a motor down and have it come to a smooth stop as a part of every move. In the case of my particular motor, it took about three revolutions for the motor to stop. Keep this in mind as we develop our programs further.

Program 15.5 **Coasting time.** Determines how many encoder counts it takes for the motor to stop.

```
CLEAR                       ; clear variables
DEFINE OSC 20               ; 20 MHz clock
DEFINE LCD_DREG PORTD       ; define lcd connections
DEFINE LCD_DBIT 4           ; 4 data bits
DEFINE LCD_BITS 4           ; data starts on bit 4
DEFINE LCD_RSREG PORTE      ; select register
DEFINE LCD_RSBIT 0          ; select bit
DEFINE LCD_EREG PORTE       ; enable register
DEFINE LCD_EBIT 1           ; select bit
LOW PORTE.2                 ; set bit low for writing to the LCD
DEFINE LCD_LINES 2          ; lines in display
DEFINE LCD_COMMANDUS 2000       ; delay in µs
DEFINE LCD_DATAUS 50        ; delay in µs
DEFINE ADC_BITS 8           ; set number of bits in result
DEFINE ADC_CLOCK 3          ; set clock source (3=rc)
DEFINE ADC_SAMPLEUS 50      ; set sampling time in µs
DEFINE CCP2_REG PORTC       ; hpwm 2 pin port
DEFINE CCP2_BIT 1           ; hpwm 2 pin bit 1
CCP1CON = %00111111         ; set status register
TRISA = %00011111           ; set status register
LATA = %00000000            ; set status register
TRISB = %00000000           ; set status register
LATB = %00000000            ; set status register
TRISC = %00000000           ; set status register
TRISD = %00000000           ; set status register
ANSEL0 = %00000001          ; page 251 of data sheet, status
                            ; register
ANSEL1 = %00000000          ; page 250 of data sheet, status
                            ; register
QEICON = %10001000          ; page 173 counter set up, status
                            ; register
INTCON = %00000000          ; set status register
INTCON2.7 = 0               ; set status register
                            ;
MOTPWR VAR WORD             ; set variables
```

(continued)

Program 15.5 Coasting time. Determines how many encoder counts it takes for the motor to stop. (*continued*)

```
PORTC.0 = 0                 ; brake off, motor control
PORTC.1 = 0                 ; PWM bit for Channel 2 of HPWM
PORTC.3 = 0                 ; dir bit for motor control
PAUSE 500                   ; LCD start up pause
LCDOUT $FE, $01, "START/CLEAR"      ; clear message
PAUSE 100                   ; pause to see message
LCDOUT $FE, $01             ; clear display
                            ;
LCDOUT $FE, $80, "GAIN =255" ; motor at full speed
HPWM 2, 255, 20000          ; C.1 PWM signal, channel 2
PAUSE 1000                  ; Let it come up to speed
HPWM 2, 0, 20000            ; C.1 PWM signal, channel 2
POSCNTH = 0                 ; set counter for encoder, H bit
POSCNTL = 0                 ; set counter for encoder, L bit
PAUSE 1000                  ; let motor stop
LCDOUT $FE, $C0, DEC3 POSCNTH," ",DEC3 POSCNTL  ;
STOP                        ; stop program to read the count
END                         ; All programs must end with END
```

Running Program 15.5 reveals that the motor spins about three revolutions once the power is shut off at full speed. The answer is 250 counts, and there are 84 counts per revolution as set up. This piece of information is important because it tells us at what rate this particular motor will stop if the motor is turned off suddenly. Any ramp down faster than that is problematic in a slow system. Store this information in the back of your mind.

In Programs 15.1 to 15.3 we implemented simple proportion and integrating gain schemes. We found that the further we are from our target, the harder the motor turns. We also found that there are times when we are only a few encoder counts from our destination, and the load is such that the motor cannot move the last few revolutions or portions of a revolution with proportional gain only. In order to overcome this situation we need an integrating function that adds to the gain if the motor is not getting to its destination over a period of time. A simple version of an integrating function was implemented in Program 15.3 (not Program 15.4!). We saw the effect of this integration by adding a little friction to the motor by holding on to the motor shaft or the encoder disc as the motor turned. Even if the motor is turned a little bit from its target position, it will slowly start to return home with a larger and larger force quickly.

CONTROLLING THE SPEED OF THE MOTOR FROM A POTENTIOMETER IN TWO WAYS

Now let us see what we need to do to use the potentiometer to control the speed of the motor. This can be done in two ways:

1. We can use the potentiometer to control the gain of the motor amplifier directly, as we did with the Program 15.4 where we determined the gain versus speed data for Table 15.1.

2. We can use what we read from the potentiometer to constantly change the target position that the controller is working toward.

The next two programs are important in that they form the basis for all motor control algorithms. Study each one carefully so you understand exactly what is going on, on each and every line of code.

First let us consider the direct control of the gain. We can use Program 15.5 to demonstrate the control of the motor speed and direction by a potentiometer. We will use a reading of 128 as the zero position and a reading on either side as positive and negative values to move the motor in either direction.

The potentiometer is controlling the motor gain and direction directly. No other changes are incorporated. The flow diagram for the algorithm is given in Figure 15.7.

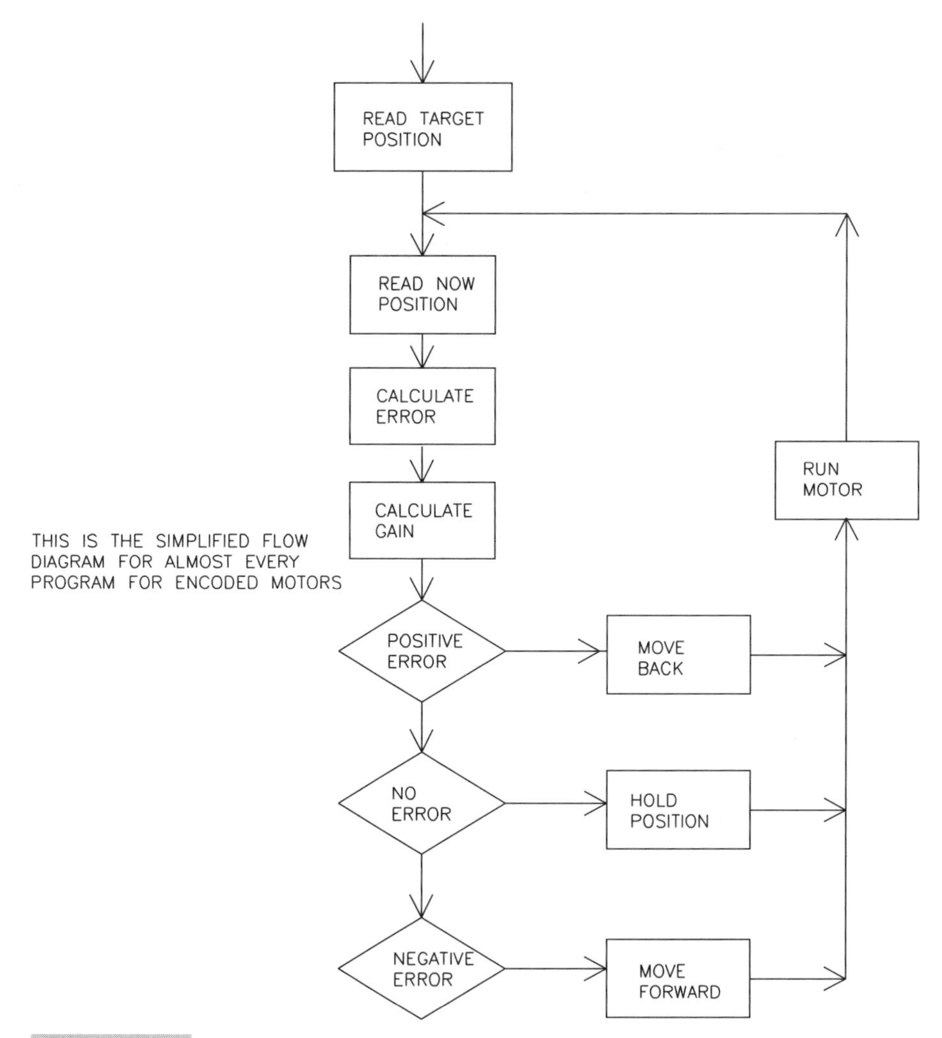

Figure 15.7 **Simplified flow diagram for typical motor position control; the input value can be from any sensor you can connect to the MCU.**

Let us first consider the simplest control that we can have over a motor. Move a motor back and forth and vary the speed with the potentiometer. We have to go from one move to next continuously. We put the reading we get from the potentiometer into the motor gain each time through the control loop, and the motor will move responding to the gain by increasing or decreasing its speed. This is implemented in Program 15.6, where a reading of 128 from the potentiometer is interpreted as the 0 speed; as we move in either direction the motor moves forward or backward. The encoder counts are not being used in the control scheme, but they are being displayed on the LCD.

Program 15.6 Controlling the speed and direction of the motor. Potentiometer reading controls the motor speed and direction directly.

```
CLEAR                      ; clear memory
DEFINE OSC 20              ; 20 MHz clock (40 not avail on the
                           ; LAB-X1)
DEFINE LCD_DREG PORTD      ; define LCD connections
DEFINE LCD_DBIT 4          ; 4 data bits
DEFINE LCD_BITS 4          ; data starts on bit 4
DEFINE LCD_RSREG PORTE     ; select register
DEFINE LCD_RSBIT 0         ; select bit
DEFINE LCD_EREG PORTE      ; enable register
DEFINE LCD_EBIT 1          ; select bit
LOW PORTE.2                ; set bit low for writing to the LCD
DEFINE LCD_LINES 2         ; lines in display
DEFINE LCD_COMMANDUS 2000     ; delay in µs
DEFINE LCD_DATAUS 50       ; delay in µs
DEFINE ADC_BITS 8          ; set number of bits in result
DEFINE ADC_CLOCK 3         ; set clock source (3=rc)
DEFINE ADC_SAMPLEUS 50     ; set sampling time in µs
DEFINE CCP2_REG PORTC      ; hpwm 2 pin port
DEFINE CCP2_BIT 1          ; hpwm 2 pin bit 1
CCP1CON = %00111111        ; set status register
TRISA = %00011111          ; set status register
LATA = %00000000           ; set status register
TRISB = %00000000          ; set status register
LATB = %00000000           ; set status register
TRISC = %00000000          ; set status register
TRISD = %00000000          ; set status register
ANSEL0 = %00000001         ; page 251 of data sheet, status
                           ; register
ANSEL1 = %00000000         ; page 250 of data sheet, status
                           ; register
QEICON = %10001000         ; page 173 counter set up, status
                           ; register
INTCON = %00000000         ; set status register
INTCON2.7 = 0              ; set status register
                           ;
```

(continued)

Program 15.6 Controlling the speed and direction of the motor. Potentiometer reading controls the motor speed and direction directly. (*continued*)

```
POSITION VAR WORD          ; set variables
MOTPWR VAR BYTE            ; set variables
POTVAL VAR BYTE            ;
SPEED VAR WORD             ;
                           ;
PORTC.0 = 0                ; brake off, motor control.
                           ; 1=brake ON
PORTC.1 = 0                ; PWM bit for Channel 2 of HPWM
PORTC.3 = 1                ; dir bit for motor control
PAUSE 500                  ; LCD start up pause
LCDOUT $FE, $01, "START UP CLEAR"   ; clear message
PAUSE 100                  ; pause to see message
POSCNTH = 127              ; set counter for encoder, H bit
POSCNTL = 0                ; set counter for encoder, L bit
                           ;
LOOP:                      ; main loop
ADCIN 0, POTVAL            ;
SELECT CASE POTVAL         ;
  CASE IS >128             ;
    PORTC.3=0              ; forward
    POTVAL=POTVAL-128      ;
  CASE IS =128             ; middle 0 position
    PORTC.3=0              ;
    POTVAL=0               ;
  CASE IS <128             ;
    PORTC.3=1              ; backward
    POTVAL=128-POTVAL      ;
  CASE ELSE                ;
END SELECT                 ;
MOTPWR=POTVAL              ;
POSITION = 256*POSCNTH + POSCNTL    ; read position registers
HPWM 2, MOTPWR, 20000      ; C.1 PWM signal actuation
COUNT PORTA.4,100,SPEED    ;
LCDOUT $FE, $80, "PORTC=",BIN4 PORTC," GAIN=",DEC3 MOTPWR
                           ; display
LCDOUT $FE, $C0, "POS=",DEC5 POSITION, " SPD=",DEC4 SPEED
                           ; display
                           ;
GOTO LOOP                  ; go back to loop
END                        ; all programs must end with END
```

In Program 15.6, we are simply controlling the gain and the motor direction. There is no motor position feedback in this program. If we add friction to the encoder wheel, the motor stops; there is no response from the gain. We are watching the four lower pins on PORTC, the gain, the motor position, and the speed.

In Program 15.7, we use the signal that we get from the potentiometer to change the target position of the motor continuously. We do this by adding the potentiometer reading to the target position of the motor. The control algorithm tries to bring the motor to its target position by working toward making the error signal zero. Code is included to take care of the position and target register overflows in each direction.

Program 15.7 Motor speed controlled by potentiometer in both directions. Value read from potentiometer added to target position continuously

```
CLEAR                        ; clear memory
DEFINE OSC 20                ; 20 MHz clock (40 better)
DEFINE LCD_DREG PORTD        ; define LCD connections
DEFINE LCD_DBIT 4            ; 4 data bits
DEFINE LCD_BITS 4            ; data starts on bit 4
DEFINE LCD_RSREG PORTE       ; select register
DEFINE LCD_RSBIT 0           ; select bit
DEFINE LCD_EREG PORTE        ; enable register
DEFINE LCD_EBIT 1            ; select bit
LOW PORTE.2                  ; set bit low for writing to the LCD
DEFINE LCD_LINES 2           ; lines in display
DEFINE LCD_COMMANDUS 2000         ; delay in µs
DEFINE LCD_DATAUS 50         ; delay in µs
DEFINE ADC_BITS 8            ; set number of bits in result
DEFINE ADC_CLOCK 3           ; set clock source (3=rc)
DEFINE ADC_SAMPLEUS 50       ; set sampling time in µs
DEFINE CCP2_REG PORTC        ; hpwm 2 pin port
DEFINE CCP2_BIT 1            ; hpwm 2 pin bit 1
CCP1CON=%00111111            ; set status register
TRISA =%00011111             ; set status register
LATA =%00000000              ; set status register
TRISB =%00000000             ; set status register
LATB =%00000000              ; set status register
TRISC =%00000000             ; set status register
TRISD =%00000000             ; set status register
ANSEL0=%00000001             ; page 251 of data sheet, status
                             ; register
ANSEL1=%00000000             ; page 250 of data sheet, status
                             ; register
QEICON=%10001000             ; page 173 counter set up, status
                             ; register
INTCON=%00000000             ; set status register
INTCON2.7=0                  ; set status register
                             ;
POSITION VAR WORD            ; set variables
MOTPWR VAR WORD              ; set variables
POTVALUE VAR BYTE            ; set variables
TARGET VAR WORD              ; set variables
MMAX VAR BYTE                ;
```

(continued)

Program 15.7 Motor speed controlled by potentiometer in both directions. Value read from potentiometer added to target position continuously (*continued*)

```
PORTC.0=0                 ; brake off, motor control. 1=brake ON
PORTC.1=0                 ; PWM bit for Channel 2 of HPWM
PORTC.3=1                 ; dir bit for motor control
PAUSE 500                 ; LCD start up pause
LCDOUT $FE, $01, "START UP CLEAR"   ; clear message
PAUSE 100                 ; pause to see message
POSCNTH=127               ; set counter for encoder, H bit
POSCNTL=0                 ; set counter for encoder, L bit
TARGET=256*POSCNTH        ;
MMAX=255                  ;
                          ;
LOOP:                     ; main loop
ADCIN 0, POTVALUE         ; read the potentiometer
POTVALUE=POTVALUE/2       ;
POSITION=256*POSCNTH + POSCNTL      ; read position registers
TARGET=TARGET+(POTVALUE-64 )  ;
PAUSE 5                   ;
SELECT CASE TARGET        ;
  CASE IS=POSITION        ;
    MOTPWR=0              ;
  CASE IS<POSITION        ; set motor direction
    PORTC.3=0             ; set direction
    PORTD.2=0             ; turn off LED
    MOTPWR=POSITION-TARGET+12  ; set motor gain
  CASE IS>POSITION        ;
    PORTC.3=1             ; set direction in reverse
    PORTD.2=1             ; turn on LED for reverse direction
    MOTPWR=TARGET- POSITION+12 ; set motor gain
  CASE ELSE               ;
END SELECT                ; end decision
IF MOTPWR>MMAX THEN MOTPWR=MMAX     ;
HPWM 2, MOTPWR, 20000     ; C.1 PWM signal
                          ;
SELECT CASE POSITION      ; Overflow prevention routine
  CASE IS<2000            ; low end
    POSCNTH=240:POSCNTL=0      ;
    TARGET=256*POSCNTH-100     ;
  CASE IS>63000           ; high end
    POSCNTH=8:POSCNTL=0 ;
    TARGET=256*POSCNTH+100     ;
  CASE ELSE               ;
END SELECT                ;
LCDOUT $FE,$80,DEC5 POSITION," PWR=",DEC3 MOTPWR," C=",BIN3
                          ; PORTC,"   "
LCDOUT $FE,$C0,DEC5 TARGET," POT=",DEC3 POTVALUE
GOTO LOOP                 ; go back to loop
END                       ; all programs must end with END
```

Program 15.7 is the first demonstration of using the counts in the encoder position register as a means of controlling the motor speed. The distance moved can now be tied to the speed, and we can design more profiles based on this information. Let's us work on this a little more.

Let us first consider the simplest move that we can make with a motor: move a certain number of encoder counts. No speed or "time to complete move" is specified. All we have to do is go from one position to the other. Under such circumstances, all we do is add the move count to the target, and the motor will move to the desired location as it works off the error signal. However, if the gain is allowed to get too high as we near the destination, the motor will go into an oscillation at its destination. As we reduce the gain, the motor oscillations will get smaller, and finally the motor will stop as desired. Program 15.8 runs the motor back and forth with a 0.5 second pause at the end of each move. It demonstrates the oscillations at high gains. Play with the potentiometer, which controls the gain, to see how the motor responds. Also note that the motor is turned off only if the error is zero when the SELECT CASE loop is at the right comparison. It is not the control situation is aggravated. This is an important realization.

Program 15.8 A simple back and forth moves of an arbitrary distance. No speed controls as such. Motor gain controlled by the potentiometer. High gains make the motor go into oscillations.

```
CLEAR                        ; clear memory
DEFINE OSC 20                ; 20 MHz clock (40 better)
DEFINE LCD_DREG PORTD        ; define LCD connections
DEFINE LCD_DBIT 4            ; 4 data bits
DEFINE LCD_BITS 4            ; data starts on bit 4
DEFINE LCD_RSREG PORTE       ; select register
DEFINE LCD_RSBIT 0           ; select bit
DEFINE LCD_EREG PORTE        ; enable register
DEFINE LCD_EBIT 1            ; select bit
LOW PORTE.2                  ; set bit low for writing to the LCD
DEFINE LCD_LINES 2           ; lines in display
DEFINE LCD_COMMANDUS 2000         ; delay in µs
DEFINE LCD_DATAUS 50         ; delay in µs
DEFINE ADC_BITS 8            ; set number of bits in result
DEFINE ADC_CLOCK 3           ; set clock source (3=rc)
DEFINE ADC_SAMPLEUS 50       ; set sampling time in µs
DEFINE CCP2_REG PORTC        ; hpwm 2 pin port
DEFINE CCP2_BIT 1            ; hpwm 2 pin bit 1
CCP1CON=%00111111            ; set status register
TRISA =%00011111             ; set status register
LATA =%00000000              ; set status register
TRISB =%00000000             ; set status register
LATB =%00000000              ; set status register
```

(continued)

Program 15.8 A simple back and forth moves of an arbitrary distance. No speed controls as such. Motor gain controlled by the potentiometer. High gains make the motor go into oscillations. (*continued*)

```
TRISC =%00000000          ; set status register
TRISD =%00000000          ; set status register
ANSEL0=%00000001          ; page 251 of data sheet, status
                          ; register
ANSEL1=%00000000          ; page 250 of data sheet, status
                          ; register
QEICON=%10001000          ; page 173 counter set up, status
                          ; register
INTCON=%00000000          ; set status register
INTCON2.7=0               ; set status register
                          ;
POSITION VAR WORD         ; set variables
MOTPWR VAR WORD           ; set variables
POTVAL VAR BYTE           ; set variables
TARGET VAR WORD           ; set variables
MMAX VAR BYTE             ;
PORTC.0=0                 ; brake off, motor control. 1=brake ON
PORTC.1=0                 ; PWM bit for Channel 2 of HPWM
PORTC.3=1                 ; dir bit for motor control
PAUSE 500 ; LCD           ; start up pause
LCDOUT $FE, $01, "START/CLEAR"      ; clear message
PAUSE 500                 ; pause to see message
POSCNTH=42                ; set counter for encoder, H bit
POSCNTL=0                 ; set counter for encoder, L bit
TARGET=256*POSCNTH        ;
SETPAUSE VAR WORD         ;
DIFFERENCE VAR WORD       ;
SETPAUSE=500              ;
DIFFERENCE=42*40          ; this is the length of the move
                          ;
LOOP:                     ; main loop
TARGET=10000              ; start here
GOSUB MAKEMOVE            ;
PAUSE SETPAUSE            ;
TARGET=10000 +DIFFERENCE      ; go here
GOSUB MAKEMOVE            ;
PAUSE SETPAUSE            ;
GOTO LOOP                 ; do it again
                          ;
MAKEMOVE:                 ;
ADCIN 0, POTVAL           ; read the potentiometer
MOTPWR=POTVAL/4           ;
POSITION=256*POSCNTH + POSCNTL      ; read position registers
```

(*continued*)

Program 15.8 A simple back and forth moves of an arbitrary distance. No speed controls as such. Motor gain controlled by the potentiometer. High gains make the motor go into oscillations. (*continued*)

```
SELECT CASE POSITION      ;
  CASE IS=TARGET          ; done
    MOTPWR=0              ; set motor power to 0
    GOSUB RUNMOTOR        ; stop
    RETURN                ;
  CASE IS<TARGET          ; under target
    PORTC.3=1             ; set direction
    GOSUB RUNMOTOR        ;
  CASE IS>TARGET          ; over target
    PORTC.3=0             ; set direction in reverse
    GOSUB RUNMOTOR        ;
  CASE ELSE               ;
END SELECT                ; end decision
GOSUB RUNMOTOR            ;
GOTO MAKEMOVE             ; go back to loop
                          ;
RUNMOTOR:                 ;
HPWM 2, MOTPWR, 20000     ; C.1 PWM signal
POSITION=256*POSCNTH + POSCNTL       ; read position registers
LCDOUT $FE, $80, DEC5 TARGET," PWR=",DEC3 MOTPWR," C=",BIN4
                          ; PORTC
LCDOUT $FE, $C0, DEC5 POSITION," POTENTMTR= ",DEC3 POTVAL
RETURN                    ;
                          ;
END                       ; all programs must end with END
```

In Program 15.8 the movement back and forth of 1680 counts work fine for small values of the potentiometer. If the values get high, the system goes into oscillations. A high gain is fine in the middle of a move but must be ramped down at the end of the move. The beginning of the move is not critical in this particular case, but that needs to be ramped too in a controlled move.

In order to make smooth moves, we need to be able to ramp the speed of the motor up and down at the beginning and end of each move. We address this problem in Program 15.9 in a preliminary way. Ramping is achieved by increasing the gain slowly on the motor at the beginning of the move and decreasing the gain slowly at the end of the move. In order to have an accurate control on the timing, we set up an interrupt routine to time the changes in the gains. The interrupt selected occurs about 100 times a second. We will ramp up for 1 second and then ramp down for 1 second.

In Program 15.9 the gain on the motor is increased one step at a time for 100 counts and then decreased one step at a time back down to zero in the next 100 counts. The program pauses and then repeats the move. You can see how far the motor has moved on the LCD. We are not using the potentiometer; the motor power is incremented and

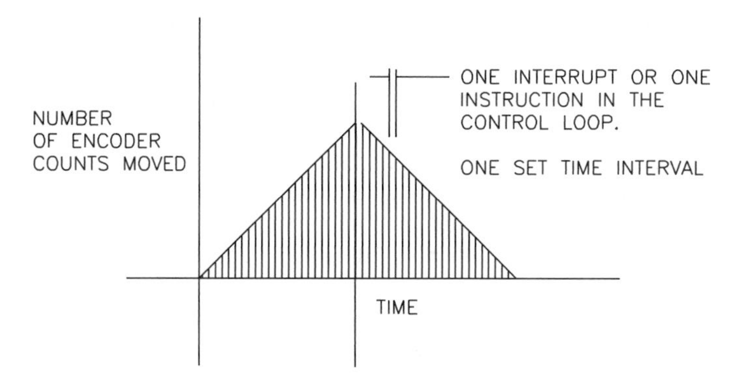

Figure 15.8 **Ramping up and down**

decremented in the interrupt routine by 1 each time through. Since we are limited to a maximum gain of 255, this has to be incorporated into the program if you decide to modify it. Figure 15.8 illustrates this in a time based graph.

Program 15.9 **Ramping up and down for 1 second each.** The gains are incremented up and down in the interrupt routine.

```
CLEAR                         ; clear memory
DEFINE OSC 20                 ; 20 MHz (40 is better if available)
DEFINE LCD_DREG PORTD         ; define lcd connections
DEFINE LCD_DBIT 4             ; 4 data bits
DEFINE LCD_BITS 4             ; data starts on bit 4
DEFINE LCD_RSREG PORTE        ; select register
DEFINE LCD_RSBIT 0            ; select bit
DEFINE LCD_EREG PORTE         ; enable register
DEFINE LCD_EBIT 1             ; select bit
LOW PORTE.2                   ; set bit low for writing to the lcd
DEFINE LCD_LINES 2            ; lines in display
DEFINE LCD_COMMANDUS 2000          ; delay in µs
DEFINE LCD_DATAUS 50          ; delay in µs
DEFINE ADC_BITS 8             ; set number of bits in result
DEFINE ADC_CLOCK 3            ; set clock source (3=rc)
DEFINE ADC_SAMPLEUS 50        ; set sampling time in µs
DEFINE CCP2_REG PORTC         ; hpwm 2 pin port
DEFINE CCP2_BIT 1             ; hpwm 2 pin bit 1
CCP1CON = %00111111           ; set status register
TRISA = %00011111             ; set status register
LATA = %00000000              ; set status register
TRISB = %00000000             ; set status register
LATB = %00000000              ; set status register
TRISC = %00000000             ; set status register
```

(continued)

Program 15.9 Ramping up and down for 1 second each. The gains are incremented up and down in the interrupt routine. (*continued*)

```
TRISD = %00000000          ; set status register
ANSEL0 = %00000001         ; page 251 of data sheet
ANSEL1 = %00000000         ; page 250 of data sheet
QEICON = %10001000         ; page 173 counter set up
INTCON = %10101100         ; set interrupt status register
INTCON2.7 = 0              ; set status register, timer0 on
T0CON = %10000000          ;
POSITION VAR WORD          ; set variables
MOT_SPD VAR WORD           ; potentiometer position
MOV_DST VAR WORD           ; potentiometer position
POTVAL VAR WORD            ; potentiometer position
MOTPWR VAR BYTE            ; motor power
INTNUM VAR WORD            ;
INTNUM = 0;                ;
MOTPWR = 0                 ;
PORTC.0 = 0                ; brake off, motor control
PORTC.1 = 0                ; PWM bit for channel 2 of hpwm
PORTC.3 = 0                ; direction bit for motor control
PORTD = 0                  ;
PAUSE 300                  ; lcd start up pause
LCDOUT $FE, $01, "START/CLEAR"      ; clear message
PAUSE 100                  ; pause to see message
POSCNTH = 0                ; set counter for encoder, h bit
POSCNTL = 0                ; set counter for encoder, l bit
ON INTERRUPT GOTO INT_ROUTINE ;
                           ;
LOOP:                      ; main loop
  POSITION=256*POSCNTH +POSCNTL      ;
  HPWM 2, MOTPWR, 20000          ; C.1 PWM signal
  GOSUB SHOW_LCD           ;
  IF MOTPWR=0 THEN         ;
    T0CON = %00000000      ;
    INTNUM = 0             ;
    HPWM 2, 0, 20000       ; C.1 PWM signal
    POSCNTH = 0            ; set counter for encoder, h bit
    POSCNTL = 0            ; set counter for encoder, l bit
    PAUSE 200              ;
    T0CON = %10000000      ;
  ENDIF                    ;
GOTO LOOP                  ; go back to loop
                           ;
SHOW_LCD:                  ; display subroutine
  LCDOUT $FE, $80, "Gain =",DEC3 MOTPWR," Cntr=",DEC3 INTNUM
  LCDOUT $FE, $C0, "POS=",DEC5 POSITION
RETURN                     ;
                           ;
```

(*continued*)

Program 15.9 Ramping up and down for 1 second each. The gains are incremented up and down in the interrupt routine. (*continued*)

```
DISABLE                 ;
INT_ROUTINE:            ; interrupt routing details
  INTNUM = INTNUM + 1   ; keep track of interrupt number
  SELECT CASE INTNUM    ; decide what to do for each interrupt
  CASE IS<=100          ; still ramping up
    MOTPWR = MOTPWR + 1 ; ramping up
  CASE IS<=200          ; now at constant speed mode
    MOTPWR = MOTPWR-1   ; ramping down
  CASE ELSE             ; f
    MOTPWR = 0          ; turn things of
  END SELECT            ;
INTCON.2 = 0            ; clear the interrupt bit
RESUME                  ;
ENABLE                  ;
END                     ; all programs must end with end
```

Program 15.9 demonstrates that we can ramp up and down quite easily by modifying the gain automatically. Next we have to do it with the distance moved.

In Program 15.10 we abandon the concept of changing the gain of the motor with the potentiometer and specify the length of the move in each step instead. (Motor counts for 2500 counts from 1000 to 3500, as shown on the LCD.) Here we are using an interrupt routine to set the distance that the motor has to move during each interrupt, and the gain is calculated by the system to reach that distance in the most efficient way possible. The motor ramps up and then ramps down and stops. The move is repeated after a short pause.

This is one of the primary lessons in motor control. Since we are telling the motor to move a little bit at a time, we must be ready with the next instruction before the previous instruction finishes executing. On the other hand if the instructions are too long or if we give the instructions too often we can also run into problems. This means that there is a critical timing sequence that has to be followed for each processor, and this timing constraint is a function of the speed of the processor and the language that we are using to control the motor. In our fairly straightforward 8-bit processor and the PBP language, we are constrained to renewing the instruction cycle about 100 times a second. This is implemented in Program 15.9. Take a look at what happens as you change the cycle timing.

In the interrupt-driven programs that come later, the interrupts are all one-hundredth of a second apart.

Program 15.10 Motor moves exactly 2500 counts in a ramp up and down mode. No adjustments are permitted in this program.

```
CLEAR                       ; clear memory
DEFINE OSC 20               ; 20 MHz clock (40 is better)
DEFINE LCD_DREG PORTD       ; define lcd connections
DEFINE LCD_DBIT 4           ; 4 data bits
```

(continued)

Program 15.10 **Motor moves exactly 2500 counts in a ramp up and down mode.** No adjustments are permitted in this program. (*continued*)

```
DEFINE LCD_BITS 4              ; data starts on bit 4
DEFINE LCD_RSREG PORTE         ; select register
DEFINE LCD_RSBIT 0             ; select bit
DEFINE LCD_EREG PORTE          ; enable register
DEFINE LCD_EBIT 1              ; select bit
LOW PORTE.2                    ; set bit low for writing to the
                               ; lcd
DEFINE LCD_LINES 2             ; lines in display
DEFINE LCD_COMMANDUS 2000      ; delay in µs
DEFINE LCD_DATAUS 50           ; delay in µs
DEFINE ADC_BITS 8              ; set number of bits in result
DEFINE ADC_CLOCK 3             ; set clock source (3=rc)
DEFINE ADC_SAMPLEUS 50         ; set sampling time in µs
DEFINE CCP2_REG PORTC          ; hpwm 2 pin port
DEFINE CCP2_BIT 1              ; hpwm 2 pin bit 1
CCP1CON = %00111111            ; set status register
TRISA = %00011111              ; set status register
LATA = %00000000               ; set status register
TRISB = %00000000              ; set status register
LATB = %00000000               ; set status register
TRISC = %00000000              ; set status register
TRISD = %00000000              ; set status register
ANSEL0 = %00000001             ; page 251 of data sheet, status
                               ; register ;
ANSEL1 = %00000000             ; page 250 of data sheet, status
                               ; register ;
QEICON = %10001000             ; page 173 counter set up, status
                               ; register;
INTCON = %10101100             ; set interrupt status register
INTCON2.7 = 0                  ; set status register
T0CON = %10000000              ;
POSITION VAR WORD              ; set variables
TARGET VAR WORD                ; set variables
ERROR VAR WORD                 ; set variable
MOTSPD VAR WORD                ; potentiometer position
MOVDST VAR WORD                ; potentiometer position
POTVAL VAR WORD                ; potentiometer position
MOTPWR VAR WORD                ; motor power
INTNUM VAR WORD                ;
COUNTER VAR WORD               ;
X VAR WORD                     ;
INTNUM = 0;                    ;
MOTPWR = 0                     ;
PORTC.0 = 0                    ; brake off, motor control
PORTC.1 = 0                    ; PWM bit for channel 2 of hpwm
PORTC.3 = 1                    ; direction bit for motor control
```

(*continued*)

Program 15.10 **Motor moves exactly 2500 counts in a ramp up and down mode.** No adjustments are permitted in this program. (*continued*)

```
PORTD = 0                     ;
PAUSE 500                     ; lcd start up pause
LCDOUT $FE, $01, "START/CLEAR"      ; clear message
PAUSE 100                     ; pause to see message
GOSUB RSTPOS                  ;
ON INTERRUPT GOTO INTROT      ; interrupt routine
                              ;
LOOP:                         ; main loop
  ADCIN 0 ,POTVAL             ; read pot value
  POTVAL=POTVAL/2             ; and divide it by 2
  POSITION=256*POSCNTH +POSCNTL      ; figure count
  GOSUB FIGURE_ERROR          ; and direction
  GOSUB RUNMOTOR              ; run motor
  GOSUB SHOW_LCD              ; display values
  IF ERROR=0 AND TARGET=3500 THEN   ; final conditions to stop
    PAUSE 1000                ; and pause
    GOSUB RSTPOS              ; reset the position target
  ENDIF                       ;
GOTO LOOP                     ; go back to loop
                              ;
FIGURE_ERROR                  :
  IF POSITION<TARGET THEN     ; not yet there
    ERROR=TARGET-POSITION     ; figure error
    PORTC.3=1                 ; set motor direction forward
  ELSE                        ;
    ERROR=POSITION-TARGET     ; figure error
    PORTC.3=0                 ; set motor direction reverse
  ENDIF                       ;
RETURN                        ;
                              ;
RUNMOTOR:                     ;
  SELECT CASE ERROR           ; decide what to do for each
                              ; INTERRUPT
    CASE IS>300               ; still ramping up
      MOTPWR=255              ; full power
    CASE IS>40                ;
      MOTPWR=ERROR            ; proportional error
    CASE ELSE                 ;
      MOTPWR=14               ; minimal power value. K++
  END SELECT                  ;
  IF MOTPWR>255 THEN MOTPWR=255      ; make sure value is not
                                     ; overflowing
  HPWM 2, MOTPWR, 20000       ; C.1 PWM signal
                              ;
  IF ERROR=0 THEN             ; at destination
    HPWM 2, 0, 20000          ; C.1 PWM signal set to 0
```

(continued)

Program 15.10 **Motor moves exactly 2500 counts in a ramp up and down mode.** No adjustments are permitted in this program. (*continued*)

```
    GOSUB SHOW_LCD          ; display values
    GOSUB FIGURE_ERROR      ; calculate error
  IF ERROR <>0 THEN         ; if not zero then run
    GOTO RUNMOTOR           ; motor to make it zero
    ENDIF                   ;
  ENDIF                     ;
RETURN                      ;
                            ;
DISABLE                     ;
INTROT:                     ; interrupt routing details
  INTNUM=INTNUM +1          ; increment counter
  SELECT CASE INTNUM        ;
    CASE IS <51             ; ramp up
      TARGET=TARGET +INTNUM ;
    CASE IS <101            ;
      TARGET=TARGET+(100-INTNUM)      ; ramp down
    CASE ELSE               ;
      T0CON=%00000000       ;reset
      INTNUM=0              ;
  END SELECT                ;
  INTCON.2 = 0              ; clear the interrupt bit
RESUME                      ;
ENABLE                      ;
                            ;
RSTPOS:                     ; reset counters
  POSCNTH = 3               ; set counter for encoder, h bit
  POSCNTL = 232             ; set counter for encoder, l bit
  POSITION=1000             ;
  TARGET=1000               ;
  T0CON=%10000000           ;
  INTNUM=0                  ;
RETURN                      ;
                            ;
SHOW_LCD:                   ; display subroutine
  LCDOUT $FE, $80, "TAR=",DEC5 TARGET, "  GAIN=",DEC3 MOTPWR
  LCDOUT $FE, $C0, "POS=",DEC5 POSITION," ERROR=",DEC3 ERROR
RETURN                      ;
                            ;
END                         ; all programs must end with end
```

In Program 15.10 the motor ramps up for 50 interrupts:

```
((50+0)/2)*50=1250
```

and then ramps down in the same 50 interrupts for a total move of:

```
1250+1250=2500
```

If the interrupts are exactly 20 microseconds apart, we have made an exact 2-second move that traversed 2500 encoder counts. The display counts from 1000 to 3500 in each move. The actual and target positions are displayed, along with the gain and the error.

SPECIFYING THE MOVE AND LETTING THE PROGRAM DO THE REST

There are certain constraints that we have to live with when we are working with a slow system and a coarse encoder. We have to tailor our instructions to the program to reflect the realities of the system that we are working with. In Program 15.10 we specify the following parameters:

- The count in the position counter that we want to start at
- The number of interrupts for the ramping up (and down)
- The rate at which we want to ramp up (and down)
- The number of interrupts we want to run for at full speed

Figure 15.9 shows us what we are going to do in graphic form and Program 15.11 implements it.

In Program 15.11, the motor starts with a count of 1000 and ramps up to from 0 to 42 counts per interrupts in 42 interrupts. It then runs at 42 counts per 100 microseconds for 200 interrupts and ramps down for 42 interrupts at one count per interrupt. The total move is 11,164 counts, and it takes 282 interrupts.

The total time for the move is the ramp up interrupts plus the run interrupts plus the ramp down interrupts. This means we can tailor the total time for the move. We can also determine the total move length from the ramping and running information. Though the results are not exact round numbers in this example, we can see what with a little work that we can get the exact move we want. We can also see that if we had

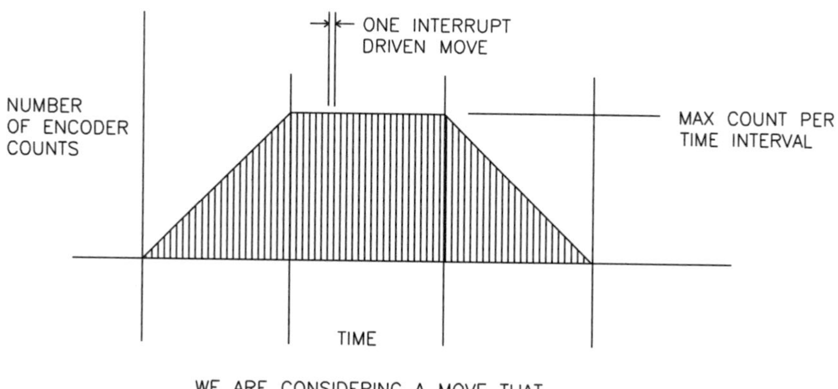

Figure 15.9 Adding ramping to a move; an interrupt-driven ramping scheme.

a higher count on our encoder our lives would be easier. Play with the program by changing the variable to see what happens. See if you can modify the program so that you can specify the move as follows:

```
The total move is to traverse exactly 10,000 counts and take
place in exactly 4 seconds. We are to ramp up for 1 second,
run for 2 seconds and ramp down for 1 second.
```

What do we have to do the get the job done? Here are some hints:

- You have to do all your reverse mathematics up front.
- You may need a slightly different ramp up and ramp down rate and distance.
- An extra number may have to be added in the run space to get an exact result.
- Never forget the usefulness of being able to see what is going on in the display.
- Learn to creep up on the problem in stages.

Program 15.11 gives you most of what you need. Improve it to get what you want.

Program 15.11 Controlled move with ramping. Ramping up, running at speed, and ramping down to specifications.

```
CLEAR                          ; clear memory
DEFINE OSC 20                  ; 20 MHz clock (40 if available)
DEFINE LCD_DREG PORTD          ; define lcd connections
DEFINE LCD_DBIT 4              ; 4 data bits
DEFINE LCD_BITS 4              ; data starts on bit 4
DEFINE LCD_RSREG PORTE         ; select register
DEFINE LCD_RSBIT 0             ; select bit
DEFINE LCD_EREG PORTE          ; enable register
DEFINE LCD_EBIT 1              ; select bit
LOW PORTE.2                    ; set bit low for writing to the lcd
DEFINE LCD_LINES 2             ; lines in display
DEFINE LCD_COMMANDUS 2000      ; delay in µs
DEFINE LCD_DATAUS 50           ; delay in µs
DEFINE ADC_BITS 8              ; set number of bits in result
DEFINE ADC_CLOCK 3             ; set clock source (3=rc)
DEFINE ADC_SAMPLEUS 50         ; set sampling time in µs
DEFINE CCP2_REG PORTC          ; hpwm 2 pin port
DEFINE CCP2_BIT 1              ; hpwm 2 pin bit 1
CCP1CON = %00111111            ; set status register
TRISA = %00011111              ; set status register
LATA = %00000000               ; set status register
TRISB = %00000000              ; set status register
LATB = %00000000               ; set status register
TRISC = %00000000              ; set status register
TRISD = %00000000              ; set status register
ANSEL0 = %00000001             ; page 251 of data sheet, status
                               ; register
```

(continued)

Program 15.11 Controlled move with ramping. Ramping up, running at speed, and ramping down to specifications. (*continued*)

```
ANSEL1 = %00000000          ; page 250 of data sheet, status
                            ; register
QEICON = %10001000          ; page 173 counter set up, status
                            ; register
INTCON = %10101100          ; set interrupt status register
INTCON2.7 = 0               ; set status register
T0CON = %10000000           ; turn on timer 0
POSITION VAR WORD           ; set variables
TARGET VAR WORD             ; set variables
ERROR VAR WORD              ; set variable
MOTPWR VAR WORD             ; motor power
INTNUM VAR WORD             ; create variable
RMPRTE VAR BYTE             ; create variable
START VAR WORD              ; create variable
RAMPUP VAR WORD             ; create variable
RUN VAR WORD                ; create variable
TTLMOV VAR WORD             ; create variable
INTNUM = 0;                 ; set variable
MOTPWR = 0                  ; set variable
PORTC.0 = 0                 ; brake off, motor control
PORTC.1 = 0                 ; PWM bit for channel 2 of hpwm
PORTC.3 = 1                 ; direction bit for motor control
PORTD = 0                   ; set variable
X=0                         ; set variable
RMPRTE=1                    ; set variable
                            ;
PAUSE 500                   ; lcd start up pause
LCDOUT $FE, $01, "START/CLEAR"      ; clear message
PAUSE 100                   ; pause to see message
ON INTERRUPT GOTO INTROT    ; interrupt routine
START=1000                  ; set variable
RAMPUP=42                   ; set variable
RUN=200                     ; set variable
GOSUB RSTPOS                ; reset the variables
LOOP:                       ; main loop
  POSITION=256*POSCNTH +POSCNTL      ; figure position
  GOSUB FIGURE_ERROR        ;and direction
  GOSUB RUNMOTOR            ;run motor
  GOSUB SHOW_LCD            ;display values
  IF ERROR=0 AND TARGET=TTLMOV THEN        ; check that move
                                           ; is done
    PAUSE 1000              ; pause to separate moves
    GOSUB RSTPOS            ; rest everything
  ENDIF                     ; end
GOTO LOOP                   ; go back to loop
                            ;
```

(*continued*)

Program 15.11 Controlled move with ramping. Ramping up, running at speed, and ramping down to specifications. (*continued*)

```
FIGURE_ERROR:                ; sub to figure error
  IF POSITION<TARGET THEN ;
    ERROR=TARGET-POSITION  ;
    PORTC.3=1              ;
  ELSE                     ;
    ERROR=POSITION-TARGET  ;
    PORTC.3=0              ;
  ENDIF                    ;
RETURN                     ;
                           ;
RUNMOTOR:                  ; run the motor
  SELECT CASE ERROR        ; decide what to do for each
                           ; interrupt
  CASE IS>300              ; still ramping up
    MOTPWR=255             ; full speed
  CASE IS>32               ; getting close
    MOTPWR=ERROR/4 +8      ; set gain
  CASE ELSE                ; last hope
    MOTPWR=16              ; minimum power to move
  END SELECT               ; end
  HPWM 2, MOTPWR, 20000    ; C.1 PWM signal
  IF ERROR=0 THEN          ; if home
  HPWM 2, 0, 20000         ; C.1 PWM signal to stop motor
  GOSUB SHOW_LCD           ; show variables
  GOSUB FIGURE_ERROR       ;
  IF ERROR <>0 THEN        ; if not home them go home
    GOTO RUNMOTOR          ;
    ELSE
    RETURN                 ;
    ENDIF                  ;
  ENDIF                    ;
RETURN                     ;
                           ;
DISABLE                    ; Required by the compiler
INTROT:                    ; interrupt routine details
  INTNUM=INTNUM +1         ;
  SELECT CASE INTNUM       ;
    CASE IS <=RAMPUP       ; start ramp up
      X=X +RMPRTE          ;
      TARGET=TARGET +X     ;
    CASE IS <=RAMPUP +RUN  ; run top speed
      TARGET=TARGET +X     ;
    CASE IS <=RAMPUP*2 +RUN   ; ramp down
      X=X-RMPRTE           ;
      TARGET=TARGET +X     ;
```

(continued)

Program 15.11 Controlled move with ramping. Ramping up, running at speed, and ramping down to specifications. (*continued*)

```
    CASE ELSE                 ;
      T0CON=%00000000         ; turn off interrupts
      TTLMOV=TARGET           ; total move
  END SELECT                  ;
  INTCON.2 = 0                ; clear the interrupt bit
  RESUME                      ; go back to program
ENABLE                        ;
                              ;
RSTPOS:                       ; set all constants
  POSCNTH = START/255         ; set counter for encoder, h bit
  POSCNTL =START//255         ; set counter for encoder, l bit
  POSITION=START              ;
  TARGET=START                ;
  INTNUM=0                    ;
  T0CON=%10000000             ; start the interrupts
  X=0                         ;
RETURN                        ;
                              ;
SHOW_LCD:                     ; display subroutine
  LCDOUT $FE, $80, "TAR=",DEC5 TARGET, "  GAIN=",DEC3 MOTPWR
  LCDOUT $FE, $C0, "POS=",DEC5 POSITION," INTRUP=",DEC3 INTNUM
RETURN                        ;
                              ;
END                           ; all programs must end with end
```

If we want to follow a complicated profile, we have to set up a lookup table that we incorporate into the interrupt routine. The interrupts have to be far enough apart to let us execute all the requirements of the lookup table between interrupts.

The motor gain can also use a lookup table in place of the SELECT CASE construct as was done in other programs.

In Program 15.12, we control the operation of the motor from an R/C radio control signal. We will use a servo exerciser to provide the radio signal equivalent. The radio signal controls the motor and creates a 1500-position remote controlled servo.

Program 15.12 Using an R/C radio signal to control the position of a motor that moves back and forth. This is position control

```
CLEAR                         ; clear memory
DEFINE OSC 20                 ; 20 MHz clock (40 better)
DEFINE LCD_DREG PORTD         ; define LCD connections
DEFINE LCD_DBIT 4             ; 4 data bits
DEFINE LCD_BITS 4             ; data starts on bit 4
DEFINE LCD_RSREG PORTE        ; select register
DEFINE LCD_RSBIT 0            ; select bit
```

(*continued*)

Program 15.12 Using an R/C radio signal to control the position of a motor that moves back and forth. (This is position control) (*continued*)

```
DEFINE LCD_EREG PORTE          ; enable register
DEFINE LCD_EBIT 1              ; select bit
LOW PORTE.2                    ; set bit low for writing to the
                               ; LCD
DEFINE LCD_LINES 2             ; lines in display
DEFINE LCD_COMMANDUS 2000      ; delay in µs
DEFINE LCD_DATAUS 50           ; delay in µs
DEFINE ADC_BITS 8              ; set number of bits in result
DEFINE ADC_CLOCK 3             ; set clock source (3=rc)
DEFINE ADC_SAMPLEUS 50         ; set sampling time in µs
DEFINE CCP2_REG PORTC          ; hpwm 2 pin port
DEFINE CCP2_BIT 1              ; hpwm 2 pin bit 1
CCP1CON=%00111111              ; set status register
TRISA =%00111111               ; set status register
LATA =%00000000                ; set status register
TRISB =%00001000               ; set status register
LATB =%00000000                ; set status register
TRISC =%00000000               ; set status register
TRISD =%00000000               ; set status register
ANSEL0=%00000001               ; page 251 of data sheet, status
                               ; register
ANSEL1=%00000000               ; page 250 of data sheet, status
                               ; register
QEICON=%10001000               ; page 173 counter set up,
                               ; status register
INTCON=%00000000               ; set status register
INTCON2.7=0                    ; set status register
                               ;
POSITION VAR WORD              ; set variables
TARGET VAR WORD                ; set variables
ERROR VAR WORD                 ;
MOTPWR VAR WORD                ; set variables
POTVAL VAR WORD                ; set variables
TOTAL VAR WORD                 ;
X VAR BYTE                     ;
PORTC.0=0                      ; break off, motor control
PORTC.1=0                      ; PWM bit for Channel 2 of HPWM
PORTC.2=0                      ; PWM bit for Channel 1 of HPWM
PORTC.3=1                      ; dir bit for motor control
PAUSE 500                      ; LCD start up pause
LCDOUT $FE, $01, "START/CLEAR"       ; clear message
PAUSE 100                      ; pause to see message
POSCNTH=125                    ; set counter for encoder, H
                               ; bit, 32000
```

(*continued*)

Program 15.12 **Using an R/C radio signal to control the position of a motor that moves back and forth.** This is position control (*continued*)

```
POSCNTL=0                        ; set counter for encoder, L bit
                                 ;
LOOP:                            ; main loop
PULSIN PORTB.3, 1, POTVAL        ;
POSITION=256*POSCNTH + POSCNTL        ; read position registers
TARGET=32000 +POTVAL/8           ;
SELECT CASE TARGET               ; find error based on target and
                                 ; position
  CASE IS=POSITION               ; at target
    ERROR=0                      ;
  CASE IS<POSITION               ; less than target
    PORTC.3=1                    ; set direction fwd
    ERROR=POSITION-TARGET        ; more than target
  CASE IS>POSITION               ;
    PORTC.3=0                    ; set direction in reverse
    ERROR=TARGET-POSITION        ; set motor gain
  CASE ELSE                      ;
END SELECT                       ; end decision
SELECT CASE ERROR                ;.
  CASE IS>50                     ;.
    MOTPWR=30                    ;.
  CASE IS>0                      ;.
    MOTPWR=14                    ;.
  CASE IS=<>0                    ;.
    MOTPWR=MOTPWR+1              ;
END SELECT                       ;.
IF MOTPWR>50 THEN MOTPWR=50      ;.
HPWM 2, MOTPWR, 20000            ; C.1 PWM signal
LCDOUT $FE, $80, "PRTC=",BIN4 PORTC,"  GAIN=",DEC3 MOTPWR
                                 ; display
LCDOUT $FE, $C0, "POS OFFSET=",DEC5 POTVAL        ; display
GOTO LOOP                        ; go back to loop
END                              ; all programs must end with END
```

USING A SERVO EXERCISER TO RUN THE MOTOR FORWARD AND BACKWARD FROM AN R/C SIGNAL (SPEED CONTROL)

A lot of the work we do with motors, as controlled remotely, has to do with running them from the signal similar to what we get from a hobby R/C receiver/transmitter. This is no different from running the motor from a PWM signal. The math we use has to convert the R/C signal to give us whatever we need for our motor. Let us take a closer look at this with the radio control signal in mind. These signals come in at pulses between 500 and 2500 microsecond with zero motion at 1500 microsecond. We want to convert this signal to a number between –127 and +127 and feed this to the HPWM instruction that will run the motor.

The pseudo code for doing this is as follows:

```
Read the signal pulse length from the radio receiver
Convert it to a number between 0 and 255
Determine the direction for the motor from the value
Subtract 127 from the the value
Feed it to the HPWM instruction
Loop to do it again
```

We can also feed the value read to the target position and use it just as we have in the earlier programs. The caveat is that the target value overflows at 0 and 65,535, and you have to reset these before this happens. In the program we are developing we will use the –127 to 127 scheme so that we can learn a new technique for doing this.

We will use pin C.7 on the MCU to receive the signal from the R/C receiver. (Any free pin can be used.) Since I did not want to broadcast a radio signal over the air, I used a servo exerciser I made with a PIC 16F819. This PIC puts out the signal I want, does not need a crystal, and is just the thing for a quick project like this. (I built the servo exerciser some time ago after destroying a number of expensive microservos by trying to turn the servo arm on them back and forth without a radio. Expensive lesson!)

See Figure 15.10 for a picture of my easy-to-make servo exerciser. Program 15.13 runs it. The circuitry is shown in Figure 15.11. This device can be programmed to create whatever electronic signals you need to experiment with your motors. Five selection jumpers on board let a large number of programs be resident at the same time. A programming header is provided, so making programming changes with the programmer and software you already have is a snap.

Program 15.13 is the program the exerciser runs on.

Figure 15.10 Servo exerciser

Program 15.13 **This program is for PIC 16F819 Servo exerciser program, only pulse lengths are implemented at this stage.**

```
CLEAR                        ; clear the memory
DEFINE OSC 4                 ; define osc speed
DEFINE ADC_BITS 8            ; set number of bits in result
DEFINE ADC_CLOCK 3           ; set internal clock source (3=rc)
DEFINE ADC_SAMPLEUS 20       ; set sampling time in µs
DEFINE CCP1_REG PORTB        ;
DEFINE CCP1_BIT 3            ;
A2D1 VAR WORD                ;
TRISA = %11110111            ;
TRISB = %11110000            ; set portb line
OSCCON = %01100000           ; 4 MHz internal
ADCON0 = %11001001           ;
ADCON1 = %01000000           ;
PORTB.0 = 0                  ; dir
PORTB.1 = 0                  ; brake
PORTB.2 = 0                  ; lock this pin down
OPTION_REG.7 = 0             ; pulls up inputs on portb
LOOP:                        ;
ADCIN 1 , A2D1               ; read incoming
HIGH PORTB.3                 ; go high
PAUSEUS 10*A2D1              ; pulse length
LOW PORTB.3                  ; go low
PAUSE 15                     ; delay 1/60 sec
GOTO LOOP                    ;
END                          ; end program
```

If we want to turn an encoded motor into an R/C controlled servo, we have to change the program so that the distance traveled is a function of the pulse width (as compared to the speed as was just done). In order to do this, we have to set up a register to represent the error signal and add and subtract the signal that we get from the radio to it. The travel distance can be increased by multiplying the signal value by a suitable number. Keep in mind that we will be limited to a discrete number of positions because we are using the PULSIN command to read the signal from the radio.

The interrupt counting program will be modified so that the increment for the motor target position is read from PORTC.7. This number can be modified mathematically in any number of ways to get the motor curve/response we want. For our immediate purposes we need to set the position for "no motor movement" first. This is the zero position for the servos and is usually represented by a pulse length of 1500 ms.

The pulse width readings from my exerciser varied from 127 to 1400. We can consider this arbitrary. In this particular case, the center reading would be 764 with a range of 637 to either side.

We will set the dormant position of the motor and the target position to 32,512 to start with. From this we will subtract the 764 middle position we got in the paragraph above. We will then read the pulse from the exerciser and add it to the target position

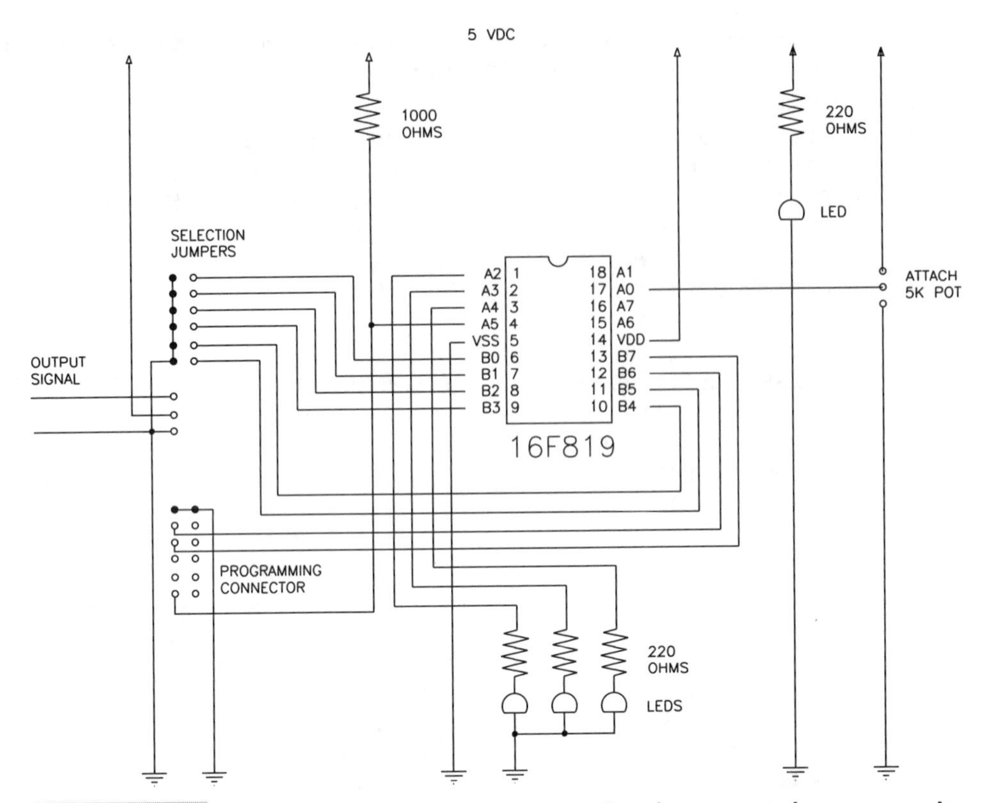

Figure 15.11 **Wiring diagram for the servo exerciser (power supply components are not shown)**

during each loop cycle. We no longer need the interrupts. A multiplier can be added to the pulse width read to extend the range of the movement.

The code to turn these potentiometer settings into a servo like operation of the motor is provided in Program 15.14.

Program 15.14 **Another program to turn servo motor into an R/C servo.** Position moved to by motor is proportional to the pulse width received.

```
CLEAR                          ; clear variables
DEFINE OSC 20                  ; 20 MHz clock (40 is better if
                               ; available)
DEFINE LCD_DREG PORTD          ; define lcd connections
DEFINE LCD_DBIT 4              ; 4 data bits
DEFINE LCD_BITS 4              ; data starts on bit 4
DEFINE LCD_RSREG PORTE         ; select register
DEFINE LCD_RSBIT 0             ; select bit
DEFINE LCD_EREG PORTE          ; enable register
```

(continued)

Program 15.14 Another program to turn servo motor into an R/C servo. Position moved to by motor is proportional to the pulse width received. (*continued*)

```
DEFINE LCD_EBIT 1            ; select bit
LOW PORTE.2                  ; set bit low for writing to
                             ; the lcd
DEFINE LCD_LINES 2           ; lines in display
DEFINE LCD_COMMANDUS 2000    ; delay in µs
DEFINE LCD_DATAUS 50         ; delay in µs
DEFINE ADC_BITS 8            ; set number of bits in result
DEFINE ADC_CLOCK 3           ; set clock source (3=rc)
DEFINE ADC_SAMPLEUS 50       ; set sampling time in µs
DEFINE CCP2_REG PORTC        ; hpwm 2 pin port
DEFINE CCP2_BIT 1            ; hpwm 2 pin bit 1
CCP1CON = %00111111          ; set status register
TRISA = %00011111            ; set status register
LATA = %00000000             ; set status register
TRISC = %10000000            ; set status register
TRISD = %00000000            ; set status register
ANSEL0 = %00000001           ; page 251 of data sheet, status
                             ; register
ANSEL1 = %00000000           ; page 250 of data sheet, status
                             ; register
QEICON = %10001000           ; page 173 counter set up, status
                             ; register
INTCON = %10101100           ; set interrupt status register
INTCON2.7 = 0                ; set status register
T0CON = %10000000            ; set timer 0
POSITION VAR WORD            ; set variables
TARGET VAR WORD              ; set variables
MOT_PWR VAR WORD             ; motor power
INT_NUM VAR WORD             ; interrupt number
MULTIPLIER var byte          ; for the pulse width read
                             ;
PORTC.0 = 0                  ; brake off, motor control
PORTC.1 = 0                  ; PWM bit for channel 2 of hpwm
PORTC.3 = 1                  ; direction bit for motor control
PAUSE 400                    ; lcd start up pause
LCDOUT $FE, $01, "START UP"   ; clear message
PAUSE 100                    ; pause to see message
POSCNTL = 0                  ; set counter for encoder, h bit
POSCNTH = 127                ; set counter for encoder, l bit
TARGET = 32515 - 764         ; No move position
MULTIPLIER = 2               ; for the pulse width read
                             ;
LOOP:                        ; main loop
PULSIN PORTC.7, 1, INT_NUM    ; read the pulse width
```

(*continued*)

Program 15.14 Another program to turn servo motor into an R/C servo. Position moved to by motor is proportional to the pulse width received. (*continued*)

```
TARGET = 32512 +INT_NUM*MULTIPLIER     ; set up the move position
  POSITION = 256*POSCNTH + POSCNTL     ; read position
  MOT_PWR = ABS(TARGET-POSITION )      ; Get absolute value
IF TARGET>POSITION THEN         ; set motor direction
  PORTC.3 = 1                   ; set direction
  PORTD.0 = 1                   ; Show direction on LED1 of
                                ; bargraph
  ELSE                          ; Decision
  PORTC.3 = 0                   ; set direction in reverse
  PORTD.0 = 0                   ; Show direction on LED1 of
                                ; bargraph
ENDIF                           ; end decision
IF MOT_PWR > 250 THEN MOT_PWR = 250   ; Limit motor power
HPWM 2, MOT_PWR, 20000          ; C.1 PWM signal
GOSUB SHOW_LCD                  ; Show display information
GOTO LOOP                       ; Go back to loop
                                ;
SHOW_LCD:                       ; Subroutine
  LCDOUT $FE, $80, "TAR=",DEC5 TARGET, " GAIN=",DEC5 MOT_PWR
  LCDOUT $FE, $C0, "POS=",DEC5 POSITION," PULSE=",DEC4 INT_NUM
RETURN                          ; Return
END                             ; All programs must end with END
```

Finally, if you want to control the speed of the motor from an R/C signal, the pulse width read from the radio has to be converted to a number between 0 and 255. This number is then used just like we used the signal from the potentiometer to run the motor. In my particular case this was done with the following code segment:

```
PULSIN PORTC.7, 1, INT_NUM    ; Read the pulse
IF INT_NUM<760 THEN        ; set motor direction
  PORTC.3 = 1              ; set direction
  PORTD.0 = 1              ; Show direction on LED1 of bargraph
  ELSE                     ; Decision
  PORTC.3 = 0              ; set direction in reverse
  PORTD.0 = 0              ; Show direction on LED1 of bargraph
ENDIF                      ; end decision
                           ;
MOT_PWR = 4 + ABS(INT_NUM-760)/3     ; get the value down to
                                     ; 255 max
IF MOT_PWR>255 THEN MOT_PWR = 255    ; Limit motor power
```

The 4 in the motor power is the allowance for the minimum power needed to move the motor. This is the friction component in the PID loop.

The entire program is listed in Program 15.15.

Program 15.15 **Another way to control the speed of an encoded motor with a hobby R/C signal.**

```
CLEAR                       ; clear variables
DEFINE OSC 20               ; 20 MHz clock (40 is better if
                            ; available)
DEFINE LCD_DREG PORTD       ; define lcd connections
DEFINE LCD_DBIT 4           ; 4 data bits
DEFINE LCD_BITS 4           ; data starts on bit 4
DEFINE LCD_RSREG PORTE      ; select register
DEFINE LCD_RSBIT 0          ; select bit
DEFINE LCD_EREG PORTE       ; enable register
DEFINE LCD_EBIT 1           ; select bit
LOW PORTE.2                 ; set bit low for writing to the lcd
DEFINE LCD_LINES 2          ; lines in display
DEFINE LCD_COMMANDUS 2000      ; delay in µs
DEFINE LCD_DATAUS 50        ; delay in µs
DEFINE ADC_BITS 8           ; set number of bits in result
DEFINE ADC_CLOCK 3          ; set clock source (3=rc)
DEFINE ADC_SAMPLEUS 50      ; set sampling time in µs
DEFINE CCP2_REG PORTC       ; hpwm 2 pin port
DEFINE CCP2_BIT 1           ; hpwm 2 pin bit 1
CCP1CON = %00111111         ; set status register
TRISA = %00011111           ; set status register
LATA = %00000000            ; set status register
TRISC = %10000000           ; set status register
TRISD = %00000000           ; set status register
ANSEL0 = %00000001          ; page 251 of data sheet, status
                            ; register
ANSEL1 = %00000000          ; page 250 of data sheet, status
                            ; register
QEICON = %10001000          ; page 173 counter set up, status
                            ; register
MOT_PWR VAR WORD            ; motor power
INT_NUM VAR WORD            ; interrupt number
PORTC.0 = 0                 ; brake off, motor control
PORTC.1 = 0                 ; PWM bit for channel 2 of hpwm
PORTC.3 = 1                 ; direction bit for motor control
PAUSE 400                   ; lcd start up pause
LCDOUT $FE, $01, "START UP"   ; clear message
PAUSE 100                   ; pause to see message
                            ;
LOOP:                       ; main loop
PULSIN PORTC.7, 1, INT_NUM    ; read the pot
IF INT_NUM<760 THEN         ; set motor direction
  PORTC.3 = 1               ; set direction
  PORTD.0 = 1               ; Show direction on LED1 of bargraph
  ELSE                      ; Decision
```

(continued)

Program 15.15 **Another way to control the speed of an encoded motor with a hobby R/C signal.** *(continued)*

```
  PORTC.3 = 0              ; set direction in reverse
  PORTD.0 = 0              ; Show direction on LED1 of bargraph
ENDIF                      ; end decision
MOT_PWR= 4 + ABS(INT_NUM-760)/3     ; set motor gain
IF MOT_PWR>255 THEN MOT_PWR = 255   ; Limit motor power
HPWM 2, MOT_PWR, 20000     ; C.1 PWM signal
GOSUB SHOW_LCD             ; Show display information
GOTO LOOP                  ; Go back to loop
                           ;
SHOW_LCD:                  ; Subroutine
  LCDOUT $FE, $80, " GAIN = ",DEC3 MOT_PWR
  LCDOUT $FE, $C0, " PULSE=",DEC4 INT_NUM
RETURN                     ; Return
END                        ; All programs must end with END
```

Other schemes can be formulated as necessary. Just follow the techniques that have been demonstrated in the preceding programs.

If you want to create a scheme in which the target is fed by the error signal you have to accommodate the condition that will underflow and overflow the two 8-bit position registers from time to time. This is done by resetting the two 8-bit position registers in the 18F4331 from time to time as they approach the upper and lower limits. The wiring diagram for controlling a motor with the 18F4331 is shown in Figure 15.12.

The wiring is the same as it would be if the PIC 18F4331 was used in a LAB-X1. As shown, the schematic has complete implementation of the programming and microswitches and so on that would be needed for a comprehensive motor control. Doing it this way allows you to put an 18F4331 in your LAB-X1 and run all the programs that we have been discussing in it or in the finished controller.

Full details of what each pin is used for as it corresponds to each of the functions in this particular layout are provided in the Appendix D on materials and suppliers, along with a photograph of a finished controller suitable for extensive experimentation. The controller shown in Figure D-1 in the appendix was used to run all the above programs for motors with encoders.

THE REALITY OF RUNNING ENCODED MOTORS

There are a number of unexpected relationships that arise when you are trying to run an encoded motor with a microprocessor. These have to do with the unfortunate fact that there is a limit to how fast things can be done with small, relatively slow microcontrollers and the language that we are using. If the motor finishes what we told it to do and we do not have the next instruction ready, the control scheme fails. The following discussion introduces you to these problems and discusses how to get around them.

The key to smooth motor operation is the routine that manages the power to the motor during the ramp up and the ramp down. What can be done in this routine is a

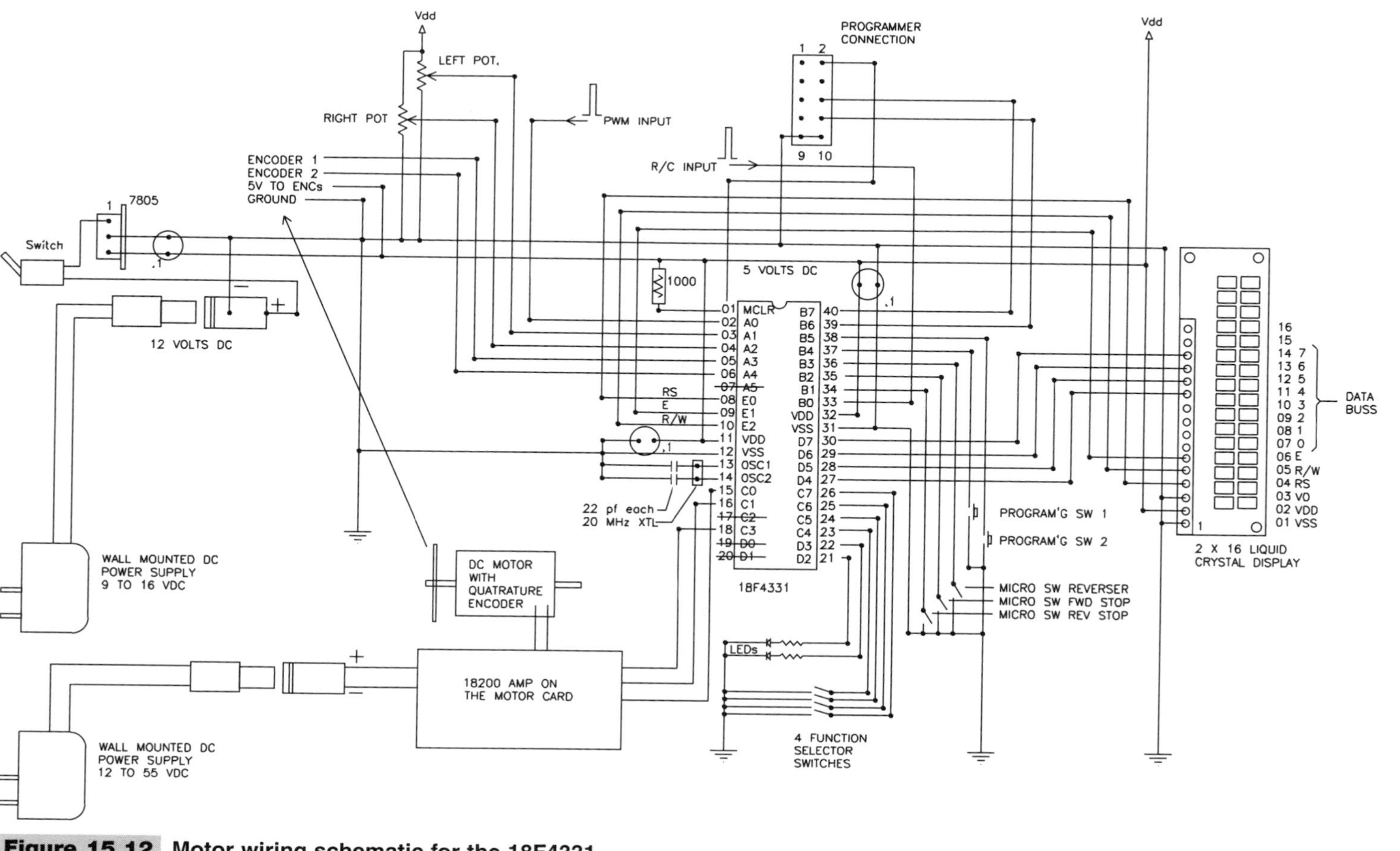

Figure 15.12 Motor wiring schematic for the 18F4331

function of how much time you have between interrupts. Select the longest interval you can tolerate and then design a detailed SELECT CASE routine to manage the power. Both the proportional and integral components need to be implemented, and the derivative function should also be implemented if there is time. The derivative function can be faked in by making the proportional and integrating values more sensitive to larger positional errors.

When you want to run a complicated "profile following program," the technique you have to use gets complicated. In the typical CNC machine, first the moves are described in the RS-274 language standard. This is the language used by all CNC machines and consists of G codes and M codes and so on, followed by *X* and *Y* positioning commands and so on. I will not go into the details of this language here but a listing of the commands and their interpretations as used by most of the FANUC systems are given in the Appendix E.

AN INTRODUCTION TO THE RS-274D LANGUAGE

In the RS-274D language, the program is interpreted as a data source that is used to create a series of pulses for as many axes as may be needed for the machine under consideration. These pulses are then fed to each axis in the operating system as needed at the appropriate rate. It is necessary to read the data base five steps ahead of the step that is being executed (for the RS 274 language); that is, it is necessary to look ahead for tool offset and interpolation considerations. The program has to look ahead to what the next instruction is to determine how far to move the machine to make sure that the current move does not compromise a future move. It turns out that up to five future moves can affect the current move. In what I have just described, we are a long way from being able to do this, and this is beyond the capability of these relatively slow MCUs; however, you should be aware of the complications.

In other words, you have to create a program that reads the first 5 lines in the RS-274D data, converts the instructions into pulses, feeds the first set of pulses to the appropriate axes for the current step, and then reads the next instruction and interprets it while the pulses are being fed to the machine. It's not simple. However, simpler schemes that will serve a one- or two-axis system may well do what we need to get done.

Note *The major difference between running a small motor and a large motor is that the larger motor needs a much more powerful amplifier to run it. A larger motor will also have a large inertial component that needs addressing in the control equation. There are many design considerations that have to be taken into account of when designing a large H bridge amplifier to control a large motor. These have to do with controlling and reversing large currents and voltages without destroying the solid state electronics and still providing automatic shut down on short circuit and internal overheating conditions. These subjects are beyond the scope of this text and do not affect your ability to control a large motor once you feel comfortable with the small ones. If you need a large amplifier, the best bet is to buy one. Its instruction sheet will give you all the information you need to get your motor running.*

16

RUNNING BIPOLAR STEPPER MOTORS

Stepper motors provide high torque at low speed and do not need an encoder to keep track of how far they have moved or of how fast they are moving. As such they are the motors of choice for office equipment and similar light duties (see Figure 16.1). In this chapter we will cover the control of the simplest form of these motors; the bipolar stepper motor. The control of other types of stepper motors uses the same techniques as we will develop next.

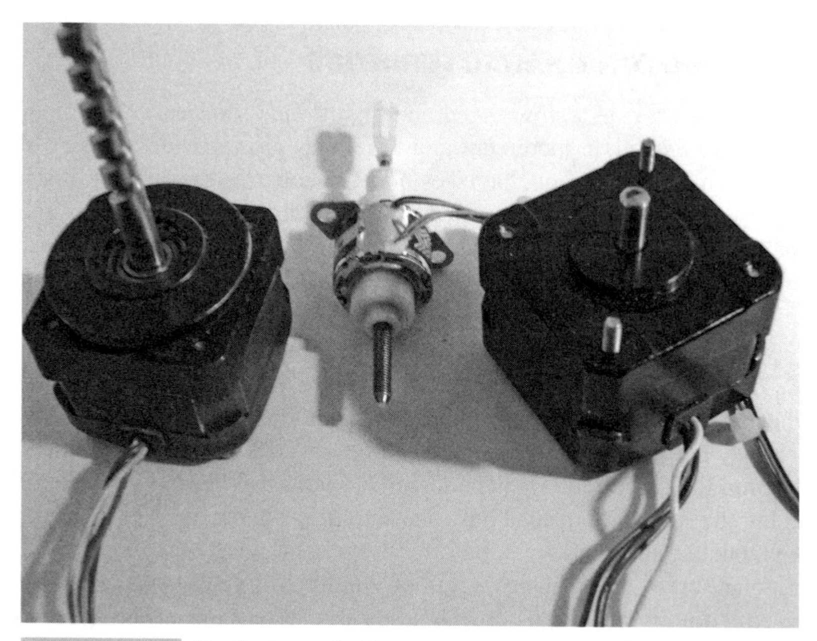

Figure 16.1 Typical small, bipolar stepper motors, which can be identified as having two coils with four wires connected to them.

Stepper Motor and Amplifier Selection

Stepper motors are made in all sorts of shapes, sizes, and configurations and with all sorts of voltage and current requirements. About the only thing they all have in common is that they move a fraction of a revolution when the electrical signals to them are moved to the next sequence. Since the techniques used to run all stepper motors are similar, we will concentrate on just one type of simple stepper motor for all our experiments (to keep costs down and the programming simple). Everything you learn about this one type of motor will be easily transferred to the running of other types of stepper motors.

The stepper motors we will be using for our experiments are small 4-wire stepper motors with only two coils inside them. These are the simplest of the stepper motors and are referred to as bipolar stepper motors. We will be using ones that need about 12 V at about 1 amp. Other stepper motors are similar in their needs; once you know how to manage this bipolar motor you should be able to run other motors without difficulty. The techniques you need to master to run this motor are the same as those that are used for all stepper motors.

Note *The larger Xavien 2-axis amplifier, discussed in Chapter 12, will handle the electrical needs of this motor with ease without need for any extra hardware and is one of the reasons for restricting ourselves to bipolar motors. Note that this amplifier needs a minimum of 12 VDC for the motor supply for its power section FETs to work properly.*

STEPPER MOTOR CHARACTERISTICS

Stepper motors provide a slow speed and high torque solution to our motion needs. Since they are moved an increment at a time, they also provide a means of keeping track of the amount of motion that takes place by counting the number of steps sent to the motor. As such the most economical way to get both speed and distance motion for small projects needing limited power is to use stepper motors if we make sure the motor is not allowed to slip.

BIPOLAR MOTORS

We selected the bipolar stepper motors because they are the simplest of the stepper motors, they are inexpensive, and they can be run with the same amplifiers we were using to run the servo motors. (The larger Xavien amplifier has two amplifiers built into it. For the servos we needed only one amplifier, but for the stepper motors we will use both, one for each coil.)

A bipolar servo motor has two sets of windings in it. When these windings are energized in the proper sequence, the motor armature rotates. The speed and direction of rotation are determined by the sequence selected and the speed at which it is executed.

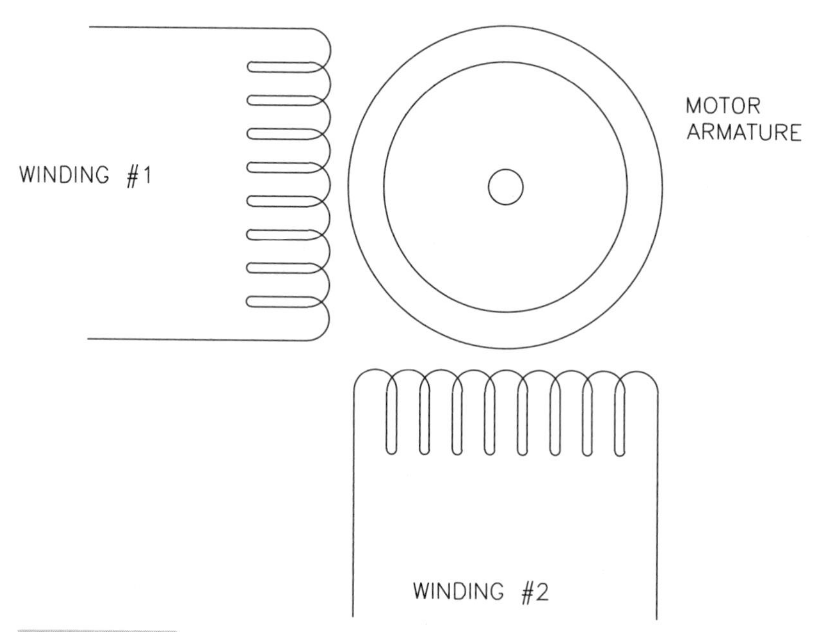

Figure 16.2 **Wiring schematic for a stepper motor—this is a schematic representation and does not represent actual coil placement positions or show the permanent magnet components.**

There are four wires on a typical bipolar motor, and they are connected to two independent windings as shown if Figure 16.2. Which two wires go together and to which winding is easily determined with a VOM. We will connect one set of windings to one of the amplifiers and the other set to the other. As long as you do not get the wiring connections mixed up, it does not matter how you connect to the amplifiers because we can reverse the polarities in the software. Even so, an orderly approach to what we do has its advantages.

Running the Motor

Once we have the motor wired to the amplifiers, we can address the business of energizing the windings in the required sequences. The sequence for movement in each direction is rigidly specified and must be followed.

There are only three things we can do as regards powering up a motor winding:

- We can send current through the winding in one direction.
- We can reverse the direction of the current.
- We can turn the current off.

We can also vary the current but this has limited utility. Sophisticated controllers use this technique to smooth the operation of the motor between steps. The technique is called microstepping; it allows intermediate spacing within the motor positions and needs a fast processor to allow all the necessary code to be executed.

The usual sequence for energizing the winding is as follows

1. Turn off all windings.
2. Turn on first winding (second winding off).
3. Turn on second winding, turn off first winding.
4. Reverse, turn on first winding, turn off second winding.
5. Reverse, turn on second winding, turn off first winding.
6. Repeat the last four steps.

The time between steps is critical. The change from one position to the other must take the same time for all steps. This is the critical part of running these motors, and this is what makes getting the motors up to high speeds without stalling difficult.

How the windings are turned on and off depend on the design of the amplifiers and what needs to be done to release one winding and energize the next. For the Xavien amplifier, the preceding steps can be expressed as the following table:

Winding 1	Winding 2
ON	Off
Off	ON
Reverse	Off
Off	Reverse
Repeat four steps	

A stepper motor can be programmed to be used in any number of ways. We will cover the following uses to determine the versatility of the motors and the ease of using them. Schemes will be developed to perform the following functions:

- Tie the speed of the motor to a potentiometer reading
- Tie the distance moved to the position of a potentiometer
- Move a motor back and forth with the following parameters:
  - Motor speed based on one potentiometer
  - Extent of motion controlled by another potentiometer

These basic techniques are the basis of all motion that most applications need. Combine them in various ways to get the results your application needs.

PROGRAMMING CONSIDERATIONS

The basic problem is to send the motor its control changes on a rigidly regular basis with a scheme that can vary the rate without losing the regularity of the changes. What are the techniques for doing this?

You *cannot* do this with the usual inline programming techniques where the program path can vary, because the time between the execution of the various instructions cannot be guaranteed and, therefore, nor can the programming path between subsequent motor power changes. This leads to an irregularity between consequent signals to the motor and thus to a choppy movement of the motor itself. Because of the harmonics that can arise in a stepper motor control scheme, this leads to problems like loss of torque and stalls at unpredictable times.

The best way to eliminate this is to create *an interrupt-based system* that provides interrupts at a strictly constant rate, where the rate can be controlled by the user with ease (with a potentiometer in our case) and at the same time at a smooth rate. The rate at which the interrupts are generated has to meet the requirements of the lowest and highest speed that the motor will be expected to operate. As a stepper motor is speeded up, its torque varies as the system goes through harmonic stages. (Search the Internet for more stepper characteristics.) It is important that the application being developed is able to work within these parameters. In the frequently seen printer application, the motor is being run at one speed that it is selected for, so this is not a problem.

Obviously, the speed at which a stepper motor operates is not continuous. It is a series of rapid steps. Because the motor moves in steps, its speed is a function of the integer stepping rate that can be executed within a time interval. Let us use 1 min as our standard time interval for now. The slowest speed for a motor under these conditions will be one step per minute. If the motor uses 200 steps per revolution, the lowest speed that the motor can be commanded to run at will be 1/200 revolutions per minute (rpm). If the maximum steps we can send it is 400 steps per second, the maximum speed will be $(400/200) \times 60$ rpm, or 120 rpm. We also need to be able to stop the motor, so the 0 speed has to be mapped within the control algorithm.

If the speed is going to be controlled from a potentiometer being read into an 8-bit byte, the 0 to 255 reading of the potentiometer has to be mapped to the 0 to 400 steps per second that the interrupt routine will generate. If the relationship can be made linear, that may be desirable for most applications.

Stepper motors also have some other characteristics that you should be aware of before we proceed further:

- There is a limit to how fast they can be accelerated. If you try to change speed too fast, they will stall.
- There are harmonic considerations within the characteristics of the motor that make their operation at certain frequencies very smooth and at other frequencies very problematic. They will also lose all torque at certain of the harmonic frequencies. Therefore, there are certain frequencies at which they cannot be run with any reliability. (The speeds at which these harmonics occur is a function of each motor and the total load on the motor.)
- The harmonic frequencies are affected by the load characteristics of the work being done, so the harmonic points can be manipulated by changing the load on the motor. Both the load and the gearing of the load can be used to manipulate the harmonic points.

- There is a limit to how fast a stepper motor can be run because of how fast we can create the interrupts needed and the rate at which the magnetic fields can be manipulated. For us this will not be a problem because under PBP it will not be possible to execute the program fast enough. However, when you change over to the much faster assembly language programming, this can be a consideration.
- The torque that a stepper motor provides varies with the speed at which it is being run and is specially sensitive to the harmonic points.

Oftentimes these handicaps can be overcome by changing the motor manufacturer or model or by changing the gearing that the motor uses to drive the load.

There are microstepping techniques that allow the motors to move to intermediate steps, but these techniques require very fast processing that can modulate the signals to the coils in real time. We will not be able to do this with the controllers and language that we are using. Smoother operation between steps is also achieved by using these techniques, especially at very slow speeds.

PROGRAMS

We will be using the 2-line LCD on the LAB-X1 to give us feedback about what is going on in our system as we run the motors. Though using the display uses up a lot of time, the benefits to us at this stage in the learning process are well worth the delays that will be caused by accessing the display constantly. We minimize the effect of the use of the LCD on the actual operation of the motor with the use of the interrupts.

First let us demonstrate the problems by running a motor without using interrupts. Let us write a simple program in which we can control the speed of the motor with a potentiometer and use the LCD to display what is being used in the way of parameters in the program.

The LCD

First let us set up the LCD display (essentially as we did in the first half of the book) and get it running. This code contains all the lines needed to completely specify what is needed. Many programs do not contain all these lines and still work because the proper conditions may be left over in the LAB-X1 from previous programming. It is always best to specify everything needed. Program 16.1 is written for the PIC 16F877A (in the LAB-X1).

Program 16.1 Making the LCD show a message. Basic LCD set up with DEFINEs

```
CLEAR                        ; Always start with clear
                             ; statement
DEFINE OSC 4                 ; Set the clock to 4 MHz
DEFINE LCD_DREG PORTD        ; Define LCD connections
DEFINE LCD_DBIT 4            ; Specify 4 bit path for the data
```

(continued)

Program 16.1 Making the LCD show a message. Basic LCD set up with DEFINEs (*continued*)

```
DEFINE LCD_RSREG PORTE       ; Port for Register select
DEFINE LCD_RSBIT 0           ; Bit for Register select
DEFINE LCD_EREG PORTE        ; Register for enable bit
DEFINE LCD_EBIT 1            ; Bit for enable bit
DEFINE LCD_RWREG PORTE       ; Define read/write register
DEFINE LCD_RWBIT 2           ; Define read/write bit
DEFINE LCD_LINES 2           ; lines in display
DEFINE LCD_COMMANDUS 2000    ; delay in µs
DEFINE LCD_DATAUS 50         ; delay in µs
DEFINE ADC_BITS 8            ; number of bits in the A to D
                             ; result
DEFINE ADC_CLOCK 3           ; clock 3 is the R/C clock
DEFINE ADC_SAMPLEUS 50       ; sample time for A to D
LOW PORTE.2                  ; Set bit low for write only
ADCON1 = %00000111           ; Set A to D control register
PAUSE 500                    ; Pause ½ second to let LCD start up
LOOP:                        ;
  LCDOUT $FE, 1, "LCD Setup OK"      ; clear Display
  PAUSE 400                  ; Pause so you can see message
  LCDOUT $FE, 1              ; clear Display again
  PAUSE 400                  ; Pause so you can see the cleared
                             ; screen
GOTO LOOP                    ; Do it forever
END                          ; End program
```

Once we have this program running we are ready to modify the LOOP by adding the commands needed to run the motors. First we will run the motor without interrupts, and then we will run it with interrupts so we can see the difference. At this stage all we are trying to do is to get the motor rotating, so what we need is a scheme to power and reverse the windings in the right sequence for moving the motor in one direction.

There are four wires on the typical bipolar motor (like the one we are using), and they are connected to two independent coils (windings).

We will be running the motor from PORTB using lines B.0 to B.5. We need three lines for each amplifier, so these six lines should do the job.

The sequence for energizing the two windings is as follows:

1. Turn off all windings.
2. Turn on second winding, turn off first winding.
3. Reverse and turn on first winding, turn off second winding.
4. Reverse and turn on second winding, turn off first winding.
5. Turn on first winding, turn off second winding.
6. Repeat the above four steps.

In table format, the sequence can be expressed as follows:

Winding 1	Winding 2
On	Off
Off	On
Reverse	Off
Off	Reverse
Repeat	

How and when the windings are turned on and off depend on the design of the amplifiers (the command structure) and what needs to be done to release one winding and energize the other.

The LCD display is set up to show what is happening on PORTB and what the value of the counter and the potentiometer are as we run the program.

```
We are using six of the lines on PORTB to control the motor
amplifiers. Lines B.7 and B.6 are not being used and are shown
as staying low in the LCD display. In order to implement the
on/off scheme shown in the preceding table, the amplifier has
to receive four signals (the other two do not change, brake
off) that are as follows on PORTB:

    00000110    The least significant bit (B.0) is on the right
                in this notation.
    00110000
    00000010
    00010000
```

The wires from PORTB to the Xavien amplifier are connected as follows:

```
Port B AMP Description
  Bit 0     Pin 1 Motor 1 Brake/Enable
  Bit 1     Pin 2 Motor 1 PWM
  Bit 2     Pin 3 Motor 1 Direction
  Bit 3     Pin 6 Motor 2 Direction
  Bit 4     Pin 7 Motor 2 Brake/Enable
  Bit 5     Pin 8 Motor 2 PWM
```

In the wiring scheme shown in Figure 16.3, the brake signal and the PWM signal can be used interchangeably, in that each one can be used as the enable signal.

```
Incorporating the commands needed to reflect the connections
that we have made, we get the following code for incorporation
into the loop:
```

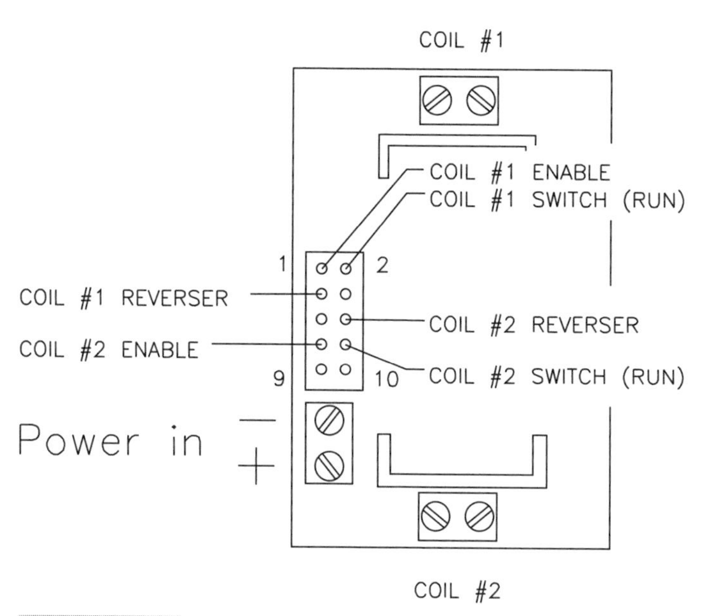

Figure 16.3 Connection points on the Xavien amplifier

```
Read the potentiometer to get P
Loop:
    PortB = %00000110
    Pause P
    PortB = %00110000
    Pause P
    PortB = %00000010
    Pause P
    PortB = %00010000
    Pause P
Goto Loop
```

The four-move loop sequence is repeated 25 times to get a total of 100 moves (steps). This is one revolution for our particular motor. The loop is repeated in the reverse direction to reverse the motor, as shown in the preceding program segment. The delay will be read from a potentiometer so that we can change it in real time.

In the pseudo program just listed, the potentiometer that controls the delay is not read in the loop. If you want to change the delay, move the potentiometer to its new location and then press the reset button on the LAB-X1 to re read the potentiometer. The new delay value will be shown on the LCD. Readjust as necessary.

The potentiometer is not read in the loop, but for each potentiometer setting the motor runs at one speed. Notice that there are discontinuities in the rotation of the motor as we run it with this interrupt-less program. The discontinuities are caused by the fact that the time between the execution of the lines in the loop is not constant. This is because the program slows down slightly when the *X* counter is reset during every fourth count. If we had read the potentiometer in the loop, the delay would have been much greater. These discontinuities will be much more apparent when the motor runs faster, and they will cause the motor to stall at a certain speed. Stepper motors require a very steady control sequence to run properly at their higher speeds. You will see this disruption as you reduce the delay to its minimum value. The motor stalls out.

Let us write the actual program, as shown in Program 16.2.

Program 16.2 Program without interrupts. Stepper motor forward and reverse 100 steps

```
CLEAR                          ; Always start with clear
DEFINE OSC 4                   ; Set the clock to 4 MHz
DEFINE LCD_DREG PORTD          ; Define LCD connections
DEFINE LCD_DBIT 4              ; Specify 4 bit path for the data
DEFINE LCD_BITS 4              ; Number of bits to be used
DEFINE LCD_RSREG PORTE         ; Port for Register select
DEFINE LCD_RSBIT 0             ; Bit for Register select
DEFINE LCD_EREG PORTE          ; Register for enable bit
DEFINE LCD_EBIT 1              ; Bit for enable bit
DEFINE LCD_RWREG PORTE         ; Define read/write register
DEFINE LCD_RWBIT 2             ; Define read/write bit
DEFINE LCD_LINES 2             ; lines in display
DEFINE LCD_COMMANDUS 2000      ; delay in µs
DEFINE LCD_DATAUS 50           ; delay in µs
DEFINE ADC_BITS 8              ; number of bits in the A to D
                               ; result
DEFINE ADC_CLOCK 3             ; clock 3 is the R/C clock
DEFINE ADC_SAMPLEUS 50         ; sample time for A to D
LOW PORTE.2                    ; Set bit low for write only
                               ; to LCD
ADCON1 = %00000111             ; Set A to D control register
PAUSE 500                      ; Pause to let LCD start up
X VAR WORD                     ; Pause variable
Y VAR BYTE                     ; Delay variable adjustment
Z VAR WORD                     ; Repeat counter
TRISA = %11111111              ; set PortA
TRISB = %00000000              ; set PortB
TRISE = %00000000              ; set PortE
  ; PORTB.0                    ; PWM (green wire)
  ; PORTB.1                    ; brake (red wire)
  ; PORTB.2                    ; dir (black wire)
LOOP:                          ; Main loop
ADCIN 0, Y                     ; Read delay variable
X = 100*Y                      ; Calculate delay
LCDOUT $FE, 1, DEC4 X," ",DEC4 y    ; Load display LCD with
                                    ; x value in dec format
```

(continued)

Program 16.2 Program without interrupts. Stepper motor forward and reverse 100 steps (*continued*)

```
FOR Z = 1 TO 25                 ; Forward 100 counts
  PORTB = %00000110             ;
  PAUSEUS X                     ;
  PORTB = %00110000             ;
  PAUSEUS X                     ;
  PORTB = %00000010             ;
  PAUSEUS X                     ;
  PORTB = %00010000             ;
  PAUSEUS X                     ;
NEXT Z                          ;
FOR Z = 1 TO 25                 ; Reverse 100 counts
  PORTB = %00000110             ;
  PAUSEUS X                     ;
  PORTB = %00010000             ;
  PAUSEUS X                     ;
  PORTB = %00000010             ;
  PAUSEUS X                     ;
  PORTB = %00110000             ;
  PAUSEUS X                     ;
NEXT Z                          ;
GOTO LOOP                       ; Repeat
END                             ; All programs end with END
```

The important thing to notice in the Program 16.2 is that there is a limit to how fast the motor can be run with this program. Notice that the motor stalls once the time between moves decreases to about 1 microsecond.

The diagram of all the wiring from the PIC16F877A to the Xavien amplifier and to the motor is shown in Figure 16.4.

Program with Regular Motor Winding Power Changes

Next let us set up a scheme to regularize the powering sequences and see how fast we can run the motor when we do it that way.

The gist of the program is the loop that changes how the windings are powered. In order to use this scheme, we have to write the requirements of what we want to put into PORTB into the EPROM part of the PIC at the beginning of the program and then read them in, one at a time as needed. The code segments for doing this are as follows:

```
Writing the EPROM
WRITE 0, %00000110      ; set how coils are energized step 1
WRITE 1, %00110000      ; set how coils are energized step 2
WRITE 2, %00000010      ; set how coils are energized step 3
WRITE 3, %00010000      ; set how coils are energized step 4
                        ;
                        ; The revised loop
                        ;
```

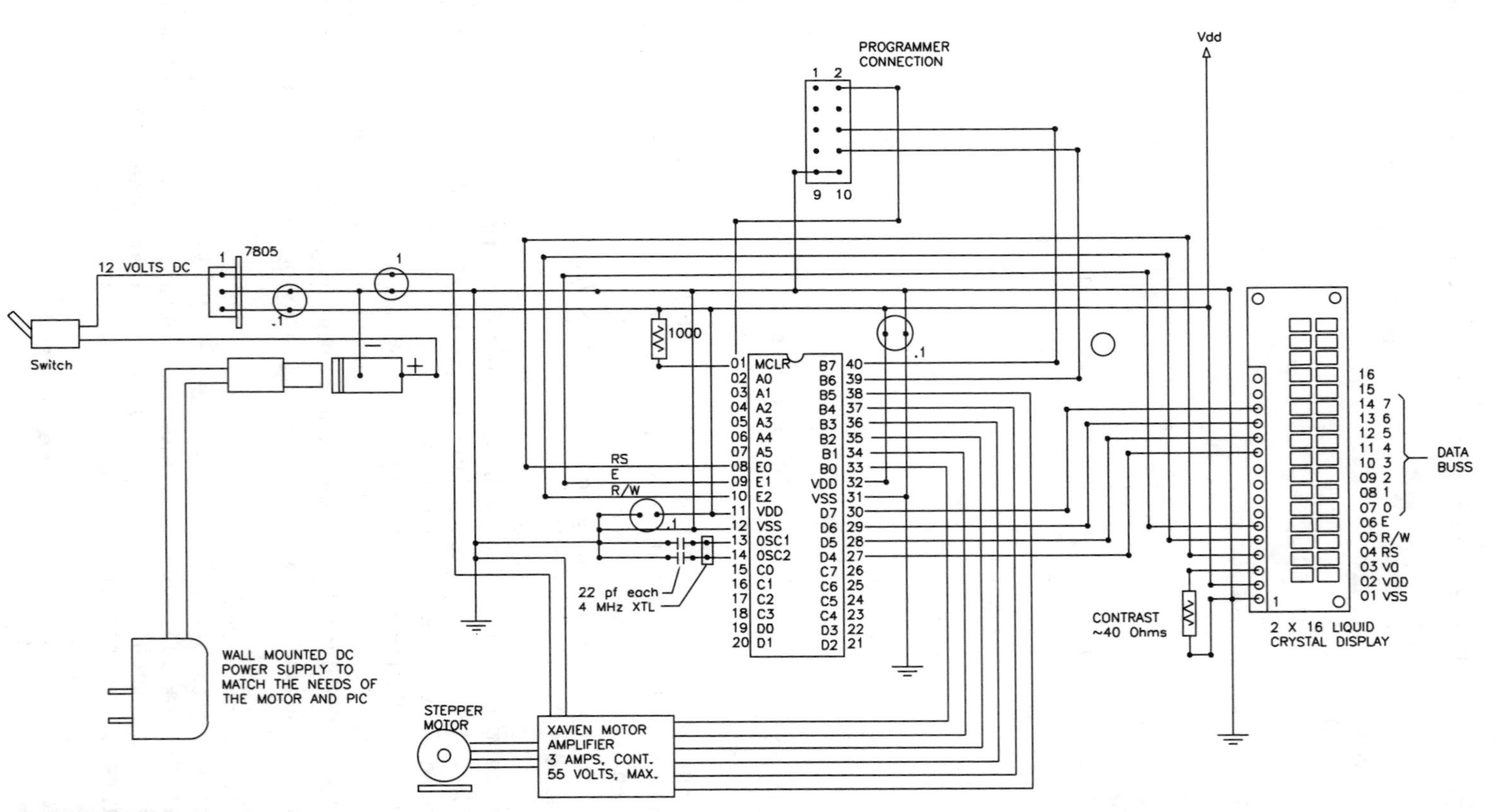

Figure 16.4 Wiring schematic for stepper motors, Xavien 2-axis amplifier, or Solarbotics amplifier

```
LOOP:                        ; Main loop
  Y = Y+1                    ; Read loop
  Y = Y & %00000011          ; picks the last two digits
  READ Y, PORTB              ; reads the right array for PortB
PAUSEUS 1000                 ;
GOTO LOOP                    ;
```

When we put this into the program we get Program 16.3.

Program 16.3 Stepper motor forward as fast as possible. Adjust the PAUSEUS 1000 constant till the motor stalls

```
CLEAR                        ; Always start with clear
Y VAR BYTE                   ;
WRITE 0, %00000110           ; set how coils are energized step 1
WRITE 1, %00110000           ; set how coils are energized step 2
WRITE 2, %00000010           ; set how coils are energized step 3
WRITE 3, %00010000           ; set how coils are energized step 4
TRISB = %00000000            ; set PortB
LOOP:                        ; Main loop
  Y = Y+1                    ; Read loop
  Y = Y & %00000011          ; picks the last two digits (0 to 3)
  READ Y, PORTB              ; reads the right bit array for PortB
  PAUSEUS 1000               ; pause should be as short as possible
GOTO LOOP                    ; do it again
END                          ; All programs end with END
```

This is one of the shortest programs in the book and will run the stepper motor as fast as possible. The PAUSEUS command in the loop should be as short as possible and still not stall the motor. For my motor this was about 1000 microseconds with the processor running at 4 MHz.

Next, we will replace the 1000 ms pause with a line to read potentiometer 0 and then set up a scheme to use the read value to modulate the speed of the motor from 0 to full speed.

The reading of the potentiometer takes about 200 ms, so we have to reduce the pause by the same amount to keep the motor running at full speed. Add the line to skip movement if the pot is at 255 and add the variable defining line for the read variable.

Making the changes gives us Program 16.4.

Program 16.4 Stepper motor speed controlled from a potentiometer

```
CLEAR                        ; Always start with clear
X VAR BYTE                   ; New variable added for pot
Y VAR BYTE                   ;
Z VAR BYTE                   ;
WRITE 0, %00000110           ; set how coils are energized step 1
WRITE 1, %00110000           ; set how coils are energized step 2
```

(continued)

Program 16.4 **Stepper motor speed controlled from a potentiometer** (*continued*)

```
WRITE 2, %00000010      ; set how coils are energized step 3
WRITE 3, %00010000      ; set how coils are energized step 4
TRISB = %00000000       ; set PortB
LOOP:                   ; Main loop
  ADCIN 0, X            ; Read delay variable
IF X = 255 THEN GOTO LOOP    ; Skip the motor winding update,
                             ; stops motor
  Y = Y+1               ; Read loop
  Y = Y & %00000011     ; picks the last two digits
  READ Y, PORTB         ; reads the right array for PortB
PAUSEUS 800 +X*50       ; Sets the delay between steps
GOTO LOOP               ; Do it forever.
END                     ; All programs end with END
```

Program 16.4 adds speed control from pot 0 on the LAB-X1, but the control is not completely linear! How would we create a linear speed profile with a potentiometer?

To run a motor really fast, we need to use assembly language programming with interrupts to control the rate at which the windings are changed. We are not covering assembly language programming in this tutorial, but we can write a program to demonstrate the techniques used. Our program will be much slower than we want, but the techniques that you need to understand will be adequately demonstrated in this program.

Program with Interrupts

Let's make sure the timer is working before things get complicated.

Here is a simple plan for confirming the operation of TIMER0:

1. We will increment the value of the variable *X* in the interrupt routine. Therefore, we will be sure that we have entered and returned from the interrupt routine if this value is being incremented.
2. We will display the value of *X* in the main loop. Therefore, if we see *X* incremented, the interrupts are being called and returned from while we are in the main loop.
3. If we see these two tasks taking place, we will have successfully used TIMER0. It's that simple. We will be ready to use TIMER0 in our programs.

Program 16.5 does this and shows the variable *X*, generated in the interrupt routine, on the LCD.

Program 16.5 **Basic interrupt routine for Timer0.** Try changing the OPTION_REG to %00000000 (pre-scaler value) and see what happens (speed)

```
CLEAR                   ; always start with clear
DEFINE OSC 4            ; define oscillator speed
```

(*continued*)

Program 16.5 **Basic interrupt routine for Timer0.** Try changing the OPTION_REG to %00000000 (pre-scaler value) and see what happens (speed) (*continued*)

```
DEFINE LCD_DREG PORTD     ; define lcd connections
DEFINE LCD_DBIT 4         ; 4 bit path
DEFINE LCD_RSREG PORTE    ; select reg
DEFINE LCD_RSBIT 0        ; select bit
DEFINE LCD_EREG PORTE     ; enable register
DEFINE LCD_EBIT 1         ; enable bit
LOW PORTE.2               ; make low for write only;
TRISD = %00000000         ; set all PORTD lines to output
TRISE = %00000000         ; set all PORTE lines to output
X VAR WORD                ; set up the variable
ADCON1 = %00000111        ; set the Analog to Digital control
                          ; register
PAUSE 500                 ; pause for LCD to start up
LCDOUT $FE, 1             ; clear screen
ON INTERRUPT GOTO INT_ROUT     ; tells program where to do on
                               ; interrupt
INTCON.5 = 1              ; sets up the interrupt enable
INTCON.2 = 0              ; clears the interrupt flag so it can
                          ; be set
OPTION_REG = %00000111    ; sets the pre-scaler to 256
X = 0                     ; sets the initial value for X
                          ;
MAIN:                     ; the main loop of the program
  LCDOUT $FE, $80, DEC5 X      ; write X to line 1
GOTO MAIN                 ; repeat to loop
                          ;
DISABLE                   ; req'd instruction, to the compiler
INT_ROUT:                 ; interrupt service routine
  X = X+1                 ; increment the X counter
  INTCON.2 = 0            ; clear the interrupt flag
RESUME                    ; go back to where you were
ENABLE                    ; req'd instruction, to the compiler
END                       ; all programs must end with End
```

Next we will modify Program 16.6 to run the stepper motor. We have to add variables, write to the EPROM, and so on and turn the motor windings on and off as needed, as we have done in previous programs. This will regularize the step sequences that the motor moves through and make for the very smooth operation that is needed to drive the motor at high speed. We will again use a potentiometer that we will read in real time to control the speed of the motor. There are 8 pre-scalers that we can apply to the timer, so we will divide the potentiometer by 32 to get a value from 0 to 7 and use that number to set the pre-scaler value. The program listing is provided in Program 16.6.

Program 16.6 Pot controlling speed via pre-scalers for Timer0

```
CLEAR                      ; always start with clear
DEFINE OSC 4               ; define oscillator speed
DEFINE LCD_DREG PORTD      ; define lcd connections
DEFINE LCD_DBIT 4          ; 4 bit path
DEFINE LCD_RSREG PORTE     ; select reg
DEFINE LCD_RSBIT 0         ; select bit
DEFINE LCD_EREG PORTE      ; enable register
DEFINE LCD_EBIT 1          ; enable bit
LOW PORTE.2                ; make low for write only;
TRISB = %00000000          ; set all PORTD lines to output
TRISD = %00000000          ; set all PORTD lines to output
TRISE = %00000000          ; set all PORTE lines to output
X  VAR WORD                ; set up the variable
Y  VAR WORD                ; set up the variable
Z  VAR WORD                ; set up the variable
WRITE 0, %00000110         ; set how coils are energized step 1
WRITE 1, %00110000         ; set how coils are energized step 2
WRITE 2, %00000010         ; set how coils are energized step 3
WRITE 3, %00010000         ; set how coils are energized step 4
ADCON1 = %00000111         ; set the Analog to Digital control
                           ; register
PAUSE 500                  ; pause for LCD to start up
LCDOUT $FE, 1              ; clear screen
ON INTERRUPT GOTO INT_ROUT      ; tells program where to do on
                                ; interrupt
INTCON.5 = 1               ; sets up the interrupt enable
INTCON.2 = 0               ; clears the interrupt flag so it can
                           ; be set
OPTION_REG = %00000111 ; sets the pre-scaler to 256
X = 0                      ; sets the initial value for X
MAIN:                      ; the main loop of the program
  LCDOUT $FE, $80, DEC3 Z      ; write X to line 1
  ADCIN 0, Z               ; read delay variable
  Z = Z/32                 ; calculate delay;
  OPTION_REG = %00000000+ Z   ; add pot reading to Opt Reg
GOTO MAIN                  ; repeat loop
DISABLE                    ; reqd instruction, to the compiler
INT_ROUT:                  ; interrupt service routine
  Y = Y+1                  ; interrupt loop
  Y = Y & %00000011        ; pick last two digits
  READ Y, PORTB            ; read port b from EPROM
  INTCON.2 = 0             ; clear the interrupt flag
RESUME                     ; go back to where you were
ENABLE                     ; reqd instruction, to the compiler
END                        ; all programs must end with End
```

In this program, only three of the pre-scaler values that were available to us would still run the motor. Others were too fast. The effect for your motor might be different.

Let us incorporate all of the preceding in our program so we can run the motor and get a hands-on idea of what this arrangement can do for us. Program 16.7 is a listing with a slightly different scheme.

Program 16.7 Running motor with Timer0 and Potentiometer 0. This program does not give us interrupts fast enough for the high speeds we want to achieve.

```
CLEAR                        ; always start with clear
DEFINE OSC 4                 ; set the clock to 4 MHz
DEFINE LCD_DREG PORTD        ; define LCD connections
DEFINE LCD_DBIT 4            ; specify 4 bit path for the data
DEFINE LCD_BITS 4            ; number of bits to be used
DEFINE LCD_RSREG PORTE       ; port for Register select
DEFINE LCD_RSBIT 0           ; bit for Register select
DEFINE LCD_EREG PORTE        ; register for enable bit
DEFINE LCD_EBIT 1            ; bit for enable bit
DEFINE LCD_RWREG PORTE       ; define read/write register
DEFINE LCD_RWBIT 2           ; define read/write bit
DEFINE LCD_LINES 2           ; lines in display
DEFINE LCD_COMMANDUS 2000        ; delay in µs
DEFINE LCD_DATAUS 50         ; delay in µs
DEFINE ADC_BITS 8            ; number of bits in the A to D
                             ; result
DEFINE ADC_CLOCK 3           ; clock 3 is the R/C clock
DEFINE ADC_SAMPLEUS 50       ; sample time for A to D
LOW PORTE.2                  ; set bit low for write only to LCD
ADCON1 = %00000111           ; set A to D control register
ON INTERRUPT GOTO INTERUPTROUTINE   ; this line needs to be
                                    ; early in the program,
                                    ; before the routine is
                                    ; called.
PAUSE 500                    ; pause to let LCD start up
X VAR WORD                   ; pause variable
Y VAR BYTE                   ; delay variable adjustment
Z VAR BYTE                   ; repeat counter
WRITE 0, %00000110           ; set how coils are energized step 1
WRITE 1, %00110000           ; set how coils are energized step 2
WRITE 2, %00000010           ; set how coils are energized step 3
WRITE 3, %00010000           ; set how coils are energized step 4
  OPTION_REG = %10000000 ; page 48 on data sheet
    ; Bit 7 =1 disable pull ups on PORTB
    ; Bit 5 =0 selects timer mode
    ; Bit 2 =0 }
    ; Bit 1 =0 } sets Timer0 pre-scaler to 1
    ; Bit 0 =0 }
```

(continued)

Program 16.7 Running motor with Timer0 and Potentiometer 0. This program does not give us interrupts fast enough for the high speeds we want to achieve. (*continued*)

```
INTCON = %10100011          ; bit 7=1 Enables all unmasked
                            ; interrupts
T1CON = %00000001           ; bit 5=1 Enables Timer0 overflow
                            ; interrupt
ADCON0 = %11000001          ; bit 2 flag will be set on interrupt
                            ; and
ADCON1 = %00000010          ;
PIE1 = %00000001            ; has to be cleared in the interrupt
                            ; routine.
                            ; It is set clear to start with.
TRISA = %11111111           ; set PortA
TRISB = %00000000           ; set PortB
TRISE = %00000000           ; set PortE
                            ;
LOOP:                       ; Main loop
  LCDOUT $FE, 1, DEC3 X, " ",DEC2 Y," ",BIN8 PORTB," ",DEC2 Z
  ADCIN 0, Z                ; read delay variable
  PAUSE 10                  ;
  Z = Z/8                   ; reduce sensitivity of constant
  GOTO LOOP                 ; repeat
DISABLE                     ; DISABLE and ENABLE must bracket the
                            ; interrupt routine
INTERUPTROUTINE:            ; this information is used by the
                            ; compiler only.
  X = X + 1                 ;
  IF X < Z THEN ENDINTERRUPT ; 1 second has not yet passed
  X = 0                     ;
  Y = Y+1                   ; Interrupt loop
  Y = Y & %00000011         ; picks the last two digits
  READ Y, PORTB             ; reads the right array for PortB
ENDINTERRUPT:               ;
  INTCON.2 = 0              ; clears the interrupt flag.
RESUME                      ; resume the main program
ENABLE                      ; DISABLE and ENABLE bracket int.
                            ; routine
END                         ; All programs end with END
```

Linear Motion: Using a Potentiometer to Position a Stepper Motor

In the previous programs we used the control signal to control the speed of the motor. Now we will use the potentiometer position to control the position of the motor. We will modify the previous programs to achieve this.

First, let us look at what it takes to move the motor back and forth with the reading from the potentiometer. Program 16.8 does this.

Program 16.8 Running motor back and forth with Potentiometer speed control

```
CLEAR                        ; always start with clear
TRISB = %00000000            ;
X VAR WORD                   ;
Y VAR WORD                   ;
Z VAR WORD                   ;
WRITE 0, %00000110           ; set how coils are energized step 1
WRITE 1, %00110000           ; set how coils are energized step 2
WRITE 2, %00000010           ; set how coils are energized step 3
WRITE 3, %00010000           ; set how coils are energized step 4
TRISB = %00000000            ; set PORTB
LOOP:                        ; main loop
ADCIN 0, X                   ; read delay variable
IF X = 255 THEN GOTO LOOP    ; skip move at this point
FOR Z = 1 TO 500             ; move 500 moves
  Y = Y+1                    ; read loop
  Y = Y & %00000011          ; picks the last two digits
  READ Y, PORTB              ; reads the right array for PORTB
  PAUSEUS 800 +X*50          ; Pause between moves, speed
NEXT Z                       ; Do it again
IF X = 255 THEN GOTO LOOP    ; Skip move at this point
FOR Z = 1 TO 500             ; move 500 moves
  Y = Y-1                    ; read loop
  Y = Y & %00000011          ; picks the last two digits
  READ Y, PORTB              ; reads the right array for PORTB
  PAUSEUS 800 +X*50          ; Pause between moves, speed
NEXT Z                       ; do it again
GOTO LOOP                    ; do it again
END                          ; all programs end with END
```

We don't need speed control for position control, so we will modify Program 16.8 to use the potentiometer reading to be the position control for the stepper.

The loop pseudo code for using the potentiometer is as follows:

```
Set the motor position as 0
Read the potentiometer
Subtract motor position from pot position
If it is 0
  Do nothing
If it is positive
  Move the motor one step in positive direction
  Increase the motor position by 1
If it is negative
  Move the motor one step in negative direction
  Decrease the motor position by 1
Go read the potentiometer again and do over
```

In Program 16.9, the LCD instructions have been added so that you can see both the position of the potentiometer and the position of the motor on the LCD.

Program 16.9 Positioning motor with the Potentiometer as position controller

```
CLEAR                        ; always start with clear
DEFINE OSC 4                 ; set the clock to 4 MHz
DEFINE LCD_DREG PORTD        ; define LCD connections
DEFINE LCD_DBIT 4            ; specify 4 bit path for the data
DEFINE LCD_BITS 4            ; number of bits to be used
DEFINE LCD_RSREG PORTE       ; port for Register select
DEFINE LCD_RSBIT 0           ; bit for Register select
DEFINE LCD_EREG PORTE        ; register for enable bit
DEFINE LCD_EBIT 1            ; bit for enable bit
DEFINE LCD_RWREG PORTE       ; define read/write register
DEFINE LCD_RWBIT 2           ; define read/write bit
DEFINE LCD_LINES 2           ; lines in display
DEFINE LCD_COMMANDUS 2000       ; delay in µs
DEFINE LCD_DATAUS 20         ; delay in µs
DEFINE ADC_BITS 8            ; number of bits in the A to D result
DEFINE ADC_CLOCK 3           ; clock 3 is the R/C clock
DEFINE ADC_SAMPLEUS 50       ; sample time for A to D
LOW PORTE.2                  ; set bit low for write only to LCD
TRISB = %00000000            ;
X VAR WORD                   ; pot reading
Y VAR WORD                   ; motor position
Z VAR WORD                   ;
WRITE 0, %00000110           ; set how coils are energized step 1
WRITE 1, %00110000           ; set how coils are energized step 2
WRITE 2, %00000010           ; set how coils are energized step 3
WRITE 3, %00010000           ; set how coils are energized step 4
TRISB = %00000000            ; set portb
ADCON1 = %00000111           ; set A to D control register
Z = 128                      ;
PAUSE 500                    ;
LOOP:                        ; main loop
LCDOUT $FE, $80, DEC4 X," ",DEC4 Z ;
  ADCIN 0, X                 ; read potentiometer variable
IF X = 128 THEN GOTO LOOP       ;
IF X>Z THEN                  ;
  Z = Z+1                    ;
  Y = Y+1                    ; read loop
  Y = Y & %00000011          ; picks the last two digits
  READ Y, PORTB              ; reads the right array for portb
  ENDIF                      ;
IF X<Z THEN                  ;
  Z = Z-1                    ;
  Y = Y-1                    ; read loop
  Y = Y & %00000011          ; picks the last two digits
  READ Y, PORTB              ; reads the right array for portb
  ENDIF                      ;
GOTO LOOP                    ;
END                          ; all programs end with END
```

Solarbotics Amplifier

If we want to use the Solarbotics amplifier for the preceding experiments, the wires from PORTB to the Solarbotics amplifier are connected as follows:

Bit 0	Pin 1	Motor 1 brake/enable
Bit 1	Pin 2	Motor 1 PWM
Bit 2	Pin 3	Motor 1 direction
Bit 3	Pin 6	Motor 2 direction
Bit 4	Pin 7	Motor 2 brake/enable
Bit 5	Pin 8	Motor 2 PWM

In the wiring scheme shown in Figure 16.5, the brake signal and the PWM signal can be used interchangeably in that each one can be used as the enable signal.

The diagram of all the wiring from the PIC16F877A to the Solarbotics amplifier and to the motor is shown in Figure 16.6. The programming does not cover the use of this amplifier, but it would be easy enough to change the control table to reflect the use of this amplifier. Everything else would remain the same.

The use of the Solarbotics amplifier is not covered in the programming examples since the programming is similar—almost identical—to the Xavien amplifier.

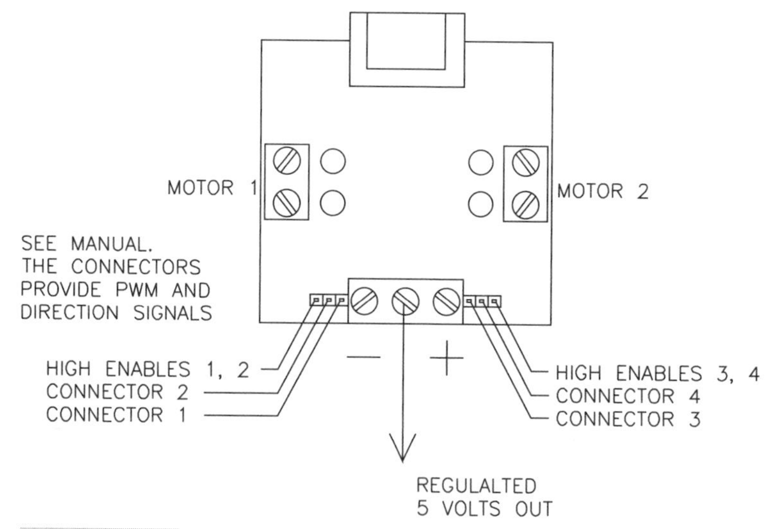

Figure 16.5 Wiring for the Solarbotics amplifier

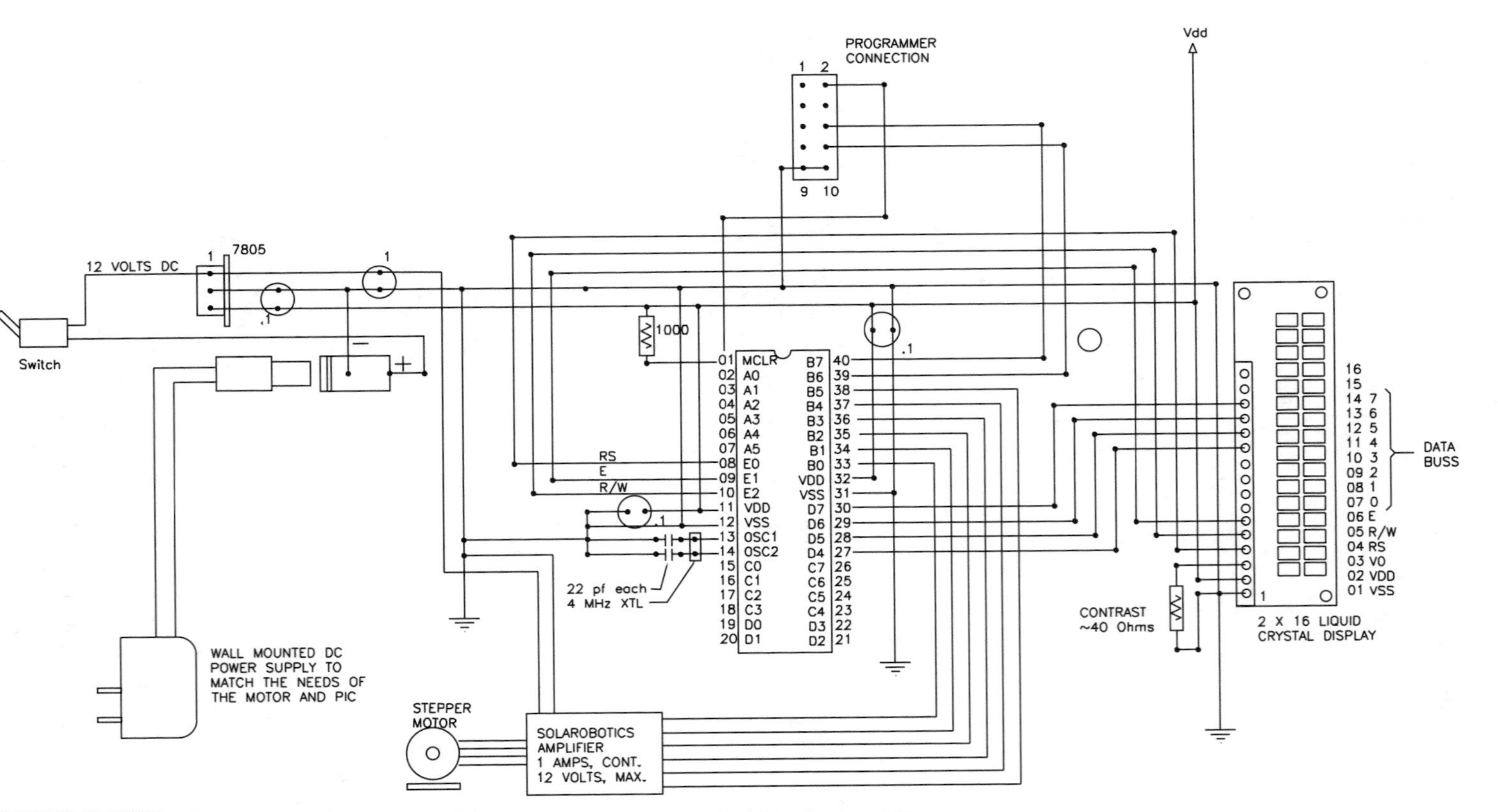

Figure 16.6 Wiring schematic for stepper motors—Solarbotics 2-axis amplifier

17

RUNNING SMALL AC MOTORS: USING SOLENOIDS AND RELAYS

There are times when it is necessary to control a small AC motor or to activate a solenoid as a part of the task we are trying to accomplish (see Figure 17.1). This chapter covers the techniques that are employed to undertake these activities.

Figure 17.1 A small fractional HP AC motor with an integral drive unit

Running a Motor

Often it is necessary for the engineer to control a small fractional horsepower electric motor as a part of his or her experimental apparatus. The easiest way to do this is to use a solid state relay that has all the components needed to allow the motor to be controlled from a TTL signal built into it. Some of these solid state relays often have the diodes needed to kill the back EMF generated when an inductive device is turned off. Relays' specifications should be checked to make sure they are suitable for the application in mind.

Solid state relays are made specifically for AC or DC applications. They may or may not have diodes built into them. Polarities have to be observed when connecting to solid state relays.

A number of vendors provide very easy to use solid state relays on the Internet. Figure 17.2 shows the relay manufactured by Crydom and available from Jameco.

Solenoids can also be run from solid state relays or from a motor amplifier or even a suitable transistor. Either the device you are using has a built-in diode, or you have to provide a diode across the solenoid to short out the back EMF generated when the solenoid is turned off. If this is not done, the solid state device can be destroyed very quickly, if not immediately. The diode is installed backward across the terminals of the solenoid so that it does not conduct when power is applied to the solenoid, as shown in Figure 17.3.

When the power to a coil is removed suddenly, the coil generates a back EMF that is proportional to the rate at which the voltage collapsed: V=dV/dt. This EMF is in the reverse direction from the original power fed to the coil. The back EMF will destroy solid state devices connected to it if it is not dissipated in some way. The diode addition provides this path and protects the solid state circuitry. Select the diode to exceed the current and voltage that was being used by a factor of from 2 to 5 to make sure the diode will not break down.

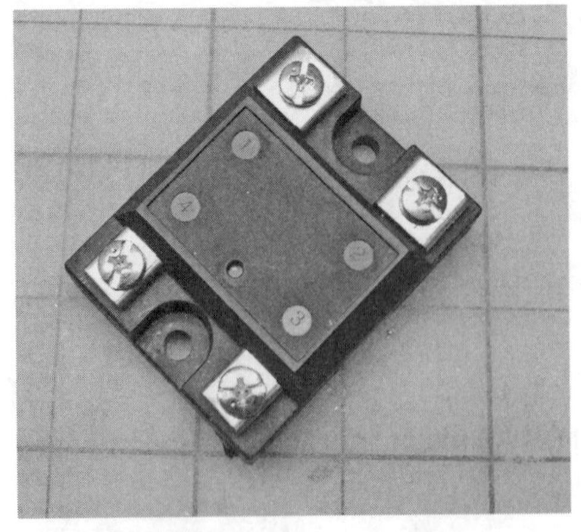

Figure 17.2 A solid state relay—note LED indicator on the relay

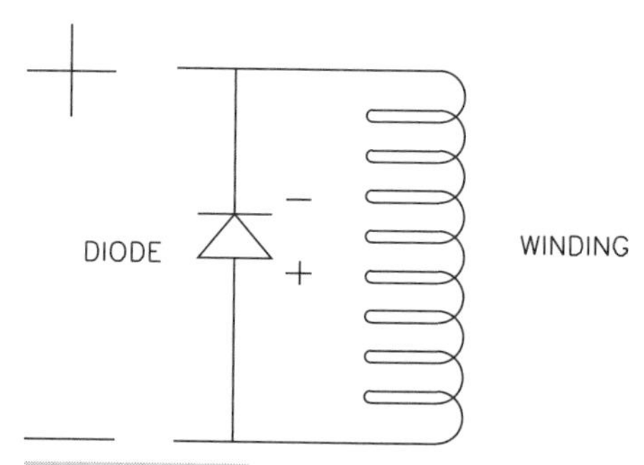

Figure 17.3 Diode polarity: how to wire a diode across a solenoid of other device with back EMF generation on disconnection

Back EMF protection must be provided if it is not already built into the device you are using. Check the specifications and the circuitry.

Some solid state relays will turn on and off when the voltage crosses the zero volts line. If this is the case for the solid state relay that you are using, you have to make sure that the controlled voltage does actually go through zero volts. This is not normally a problem for AC loads.

Standard off the shelf solid state relays cannot be used in PWM mode. As a matter of fact, these devices are best turned off and on every few seconds or more. The usual AC motor too does not do well if turned on and off more than a few times a minute. It takes these motors a few seconds to come up to speed if they are cycled on and off too often; they will overheat. You can also damage the centrifugal switch, used on the motors that use a separate starting and running winding, if it is cycled constantly.

Solid state relays should not be used to control the speed of an AC motor or an AC/DC motor. Much faster switching than they are designed for is needed.

Using a Relay

If you have a difficult situation where you are not sure what might be the best way to turn a device on and off, consider using an intermediate relay to serve as your isolation device. Use the solid state relay to turn on the mechanical relay, and use the mechanical relay to turn on the motor (or whatever you need). Not only is this a good isolation procedure, but it allows you to control large loads with a small solid state device.

You have to place a diode across the relay windings, as discussed earlier in this chapter.

A solid state relay can also be used to pull in the coil in the starter relay for a large motor. Be sure to match the voltage and amperage requirements of the starter coil. If there are interlocks within the starter relay control circuitry, these will have to be accommodated.

18

DEBUGGING AND TROUBLESHOOTING

Debugging is not a random process during which one might hope to get lucky. It is a very carefully thought out strategy to find out why a circuit or program is not behaving the way it was intended to and what needs to be done to correct the problem. You will have fixed the problem only if you can make the problem come back by undoing the fix. A vague superstition that you might have fixed the problem by pressing on a warm resistor is not enough to decide that the problem has been solved. You must be able to make the problem come back. This is exactly the reason why intermittent problems are so hard to fix. It's hard to make the problem come and go on command, and so it is also harder to understand what the problem is.

Problem: The Microcontroller Crystal Circuit Must Oscillate

If the PIC oscillator will not oscillate, nothing can be done to fix anything, so the first thing we have to do is make sure the OSC lines are actually oscillating. The easiest way to check this is with an oscilloscope. If there are problems, the following points are relevant:

- The LAB-X1 runs at 4 MHz out of the box. Your system should also be set to run at 4 MHz so that you can use the LAB-X1 as a test bed whenever you need to. This allows you to take the PIC back and forth between your project and the LAB-X1 to see where the problems are.
- To start with, do all your projects with a 4 MHz setting. Later you can move to 20 MHz.
- If the crystal frequency does not match the DEFINE OSC statement in your program, there will be problems. If you are using a crystal or resonator, you must know what the frequency of the device is, and your software must state this

number in the OSC statement at the top of the program. The PIC manages the power to the oscillator based on this information, and if this is incorrect the oscillator will not have the right amount of power to oscillate freely. On your circuitry where you are not using a dividing network as is done on the LAB-X1, the values of the capacitors at the crystal are critical. They must be of the value recommended for the crystal you are using.

- The Configuration and Option pull-down menus in the programmer must match the actual conditions in the hardware, and they must match each other.

HARDWARE CHECKS

Start off with the following checks on the hardware side:

1. Make sure the microcontroller has power:
   - Make sure it is 5 V on the money.
   - Make sure there is power to both sides of the microcontroller if that is required.
   - Use an oscilloscope so that you can see what is going on.
2. Make sure the MCLR pin has been pulled up to 5 V with an appropriate (10K) resistor:
   - Make sure that the MCLR pin is actually high. Put a meter on it. Use a scope.
3. Make sure that the oscillator is running. Use an oscilloscope. Make sure the operation is clean and consistent.
4. Use your eyes to check the PC board for shorts and dry solder joints. Use a magnifying glass. Go over questionable areas with a soldering iron again and recheck your work.
5. Make sure the wiring is what it is supposed to be. Check the route of every wire. Mark it off on the schematic as you check it. Check the PC board trace routings where necessary. Recheck your design to make sure you got it right. Have a friend to check it.
6. Check the values of each of the components on the board.
7. Make sure each IC is oriented with pin 1 in the proper location in its socket or on the PC board.
8. Make sure power and ground to each IC are properly routed and are actually power and ground.
9. Measure voltages throughout the layout and confirm that they are what they are supposed to be.
10. Make sure all capacitors are installed correctly. Check polarity where necessary.
11. Make sure all diodes are installed correctly. Not all always need to be installed so that the cathode is connected to ground. Confirm connections on all inductive loads.
12. Make sure all inductive loads are properly protected against with diodes. Make sure these diodes are properly rated for amperage, voltage, and switching speed.
13. Use the oscilloscope to check for noise in the circuitry. Eliminate it by adding small capacitors at the noisy areas.

14. Read the relevant pages of the data sheet (on the startup pages) again. Make sure you understand what the data sheet is saying.
15. Make sure that the 10-wire cable between the programmer and the PIC board is making proper contact on all 10 pins.

SOFTWARE CHECKS

Make the following checks on the software side:

1. Go over the software line by line and make sure there are no typing mistakes. Not all typing mistakes are identified by the compiler software.
2. Write a short LED blink routine and run that on the board to make sure the system is actually alive and working. If the program loops as a part of its design, add a blinking instruction for one of the LEDs to tell you that the loop is actually executing as designed and not hanging up on some segment of code somewhere else in the program.
3. Use MicroCode Studio to see what is going on in the system as it is run. The software is free for the unloading off the Internet for individual users and will run the 16F877A. Details on using this or other similar software are beyond the scope of this book and are not covered herein.
4. For step 3 you can set up your PC as a dumb terminal and connect it to the board so that the entire system can be viewed on the screen with appropriate software.
5. Go over the software to make sure there are no logical mistakes.
6. Follow the use of each variable throughout the program to make sure that it does not get modified where is not supposed to be modified.
   The software we are using uses integer math and 8 and 8/16-bit variables. Make sure that none of the variables exceed the bounds that have been designated by the variable sizes. Results with negative answers are problematic in integer math. The minus sign is not implemented in the results!
7. Read the relevant pages of the data sheet again for the part of the MCU that you are having problems with. Make sure you understand what the data sheet is saying.

FEEDBACK

There are a number of ways to get input into and feedback out from a malfunctioning program. Using as many of these methods as possible will reduce the number of times you have to reload the program. The following feedback devices are available to you. Incorporate what you need into critical areas of the program.

1. **LCD display** Use both line 1 and line 2 and use each character on each line.
2. **The speaker** It's easy to set up two tones that are easy to differentiate.
3. **The dumb terminal** It can display a large amount of information. Use the DEBUG command in PBP to send information to a dumb terminal. This command is explained in detail in the PBP manual.

4. **The various buttons** Use these to modify what is happening in the program in real time.
5. **The three on board potentiometers** Use these to input various values into the program and see how they modify its operation.
6. **Add LEDs** If more are needed, add these to the hardware so you can turn them on and off at critical junctures to confirm what you have programmed.

You can also insert a short loop that displays the registers that you are interested in at a critical location in the program. When the program enters this loop, it indicates that the program actually got this far and then it displays the registers of interest again and again without going any further. This loop can be moved up and down through the program to see what is going on where.

Programming problems can be also be tracked down with the following simple procedures. You should have a similar list of procedures that you have developed and that you are comfortable with for troubleshooting all your own work.

1. Determine if the program is actually getting to a certain critical line of code.
2. Determine what the contents of various registers are at critical times in the code.
3. Look at how counters are behaving and confirm to yourself that this is exactly what is supposed to be happening.
4. Display data based on interrupts programmed and/or entered by you from the keys.
5. Look for areas where the program might be getting stuck in a loop.
6. Pay special attention to the handling of interrupts.
7. Go over the circuit layout to make sure there are no mistakes in the design of the circuitry.
8. Go over the physical circuit to make sure it is actually wired the way it was designed.
9. Make sure that all lines that are to be pulled up or down are actually being pulled up and down and that the resistors being used are of the right values.

If the circuitry refuses to run at all, check the configuration settings in the programmer software. The oscillator configuration is the most critical, but other settings can also prevent the PICmicro from starting up. See the "Special Features of the CPU" pages in the data sheet for correct configuration details. The default settings for configuring the various conditions are discussed later, near the end of this chapter.

Using the PBP Compiler Commands to Help Debug a Program

The PICBASIC PRO compiler provides a number of commands that can be a tremendous help in debugging programs that refuse to cooperate. These commands can be broken down into a number of categories to better understand how they can be used.

COMMANDS THAT CAN PROVIDE DEBUG OUTPUT TO A SERIAL PORT

A number of the commands provided by the compiler do not have a function other than to aid in the debugging of programs by outputting data to a designated pin on the microcontroller. This data can then be displayed on a dumb terminal also.

The following commands are useful in the context (see the PBP book):

```
DEBUG is like a print command to the serial port (and so
to the dumb terminal).
DEBUGIN
ENABLE
ENABLE DEBUG
DISABLE
DISABLE DEBUG
PEEK
POKE
SOUND
```

DUMB TERMINAL PROGRAM

There are a number of terminal programs that are available at no charge on the Internet. I use the dumb terminal program provided by Microsoft as a part of their operating system utilities. It provides all the functionality that you need to use it with the PRO Basic Compiler and MicroCode development software.

THE BRAY TERMINAL PROGRAM

This is a more sophisticated dumb terminal program also available free on the Internet.

SOLDERLESS BREAD BOARDS

Using solderless boards for your prototyping activities is, in general, not recommended. They are all right for small experimental excursions when you first start out with the microcontrollers, but as your circuitry gets more and more sophisticated and complicated, there is too much of a probability for poor connections and wires that come loose to use these devices.

It is recommended that you use the perforated boards that have a separate solder pad for each hole, and then solder each component into the board. Then wire each of the components with hook up or wire, wrap wire with straightforward point to point wiring. The key is to be very careful and thorough so that there are no mistakes. It takes patience and care. Take your time and check your work before and after each connection is made.

I also use circuit boards with continuous bars of conductors on them.

Debugging at the Practical Level

"Now what do I do? The project is deader than a doornail and I don't have a clue!"

A fairly long program that you wrote will not work the way it is supposed to. You don't know if it is the software or the hardware, and you do not have a clue, what you should do? Don't throw it all in the garbage just yet. Chances are that with a little bit of work everything will be just as you intended. After all, you did create all this code.

The problem is that there is nothing to look at or to see, the thing is dead, and you don't know where to start. The solution is to make things visible and to start the process in a step by step manner so that you can make sure that each step in the program you created is doing what it is supposed to do. The good news is that you do not have to spend a fortune on new software and hardware, and you don't have to spend a year of your life learning a new discipline. You already know and have everything you need to debug the program in your LAB-X1 board.

There are three output devices on the LAB-X1 board that can be used as aids in the debugging process:

- The LCD display
- The 8 LEDs in the LED bar
- The piezoelectric speaker device

There are also a number of input devices that can aid in the debugging process by making it possible to make the debugging more interactive:

- The keypad
- The three potentiometers
- The reset button!

We also have some of the standard software tools that we can use:

- The PAUSE command
- The STOP command
- The IF..THEN couplet
- The ability to COMMENT out sections of code

The PICBASIC PRO compiler provides a number of statements that are designed specifically for the debugging process. These are mentioned previously and should be studied in the PBP book.

The personal computer that you are using is also a powerful debugging tool in that it can both send and receive information and gives you a full screen and a keyboard to use as interactive elements. PICBASIC PRO provides a number of powerful tools to let you interact with your PC. However, the most powerful device at your command is the computer between your ears. By and large, the debugging process is an exercise in the use of the brain. Everything else that needs to be done can be done with the LAB-X1 and your personal computer.

There are some rules that you need to follow that will make the debugging process easier and more likely to succeed in a reasonable time:

Rule 1 Be thinking about the debugging process as you write the code. Design the code so that it can be debugged and put in the necessary hooks and connects as you go along.

Rule 2 Write the code as small subroutines that can be tested as standalone mini-programs. Once you have the software working you can streamline the code. Test your program as you develop it to make sure each developmental level is operational.

Rule 3 Do not wait till the last moment to start the debugging process. Debug as you go along, meaning debug the code as it is developed rather than waiting till it is all done and ready to be delivered to the customer. Learn to write the code so that you can debug sections of code as sections of code or as standalone subroutines.

Rule 4 Write a set of routines that can be called from within the code that shows you what the content of various memory locations on the LCD or the bar graph as the program runs.

The first thing most programs must do is make the LCD come alive. There are three things that have to be checked to confirm its proper operation. Is the software right? Are the DEFINEs correctly called? Have you some how destroyed the electronics?

SOFTWARE

PBP (PICBASIC PRO) makes it completely painless to use the LCD. All you have to do it to tell the software where the LCD is connected to the hardware and which pins are connected to what function on the LCD. In the case of the LAB-X1 board that we are using, the LCD is connected as follows:

- The LCD is connected to Port D and Port E.
- It can use the 4-bit or 8-bit mode to send data to the LCD (if all pins are connected).
- The Register select bit is at Port E bit 0.
- The Enable bit is at Port E bit 1.
- The Read/Write bit is Port E bit 2 and is made low for writing.

These variables are defined by the following code segment. This code should be placed at or near the beginning of your program:

```
DEFINE LCD_DREG     PORTD   ; LCD connected to PORTD
DEFINE LCD_DBIT     4       ; uses 4 bit data path
DEFINE LCD_RSREG    PORTE   ; RESET register is PORTE
DEFINE LCD_RSBIT    0       ; uses bit 0
DEFINE LCD_EREG     PORTE   ; ENABLE register is PORTE
DEFINE LCD_EBIT     1       ; uses bit 1
PORTE.2 = LOW               ; we will be writing only
```

You *must* have a pause of about 0.5 seconds at the start of your program, before you first access the LCD, to allow the LCD to complete its setup routines. If this pause is omitted, the LCD can malfunction or may not start up at all. Just to make sure, start with a PAUSE of 0.5 seconds and then shorten it when you know that everything is working properly:

```
PAUSE 500     ; and
ADCON1 = %00000111      ; set the digital modes needed. This
                        ; has to do with
              ; making PORTE digital for controlling the LCD.
              ; PORTE is analog on startup and reset. PORTD
              ; is digital only and cannot be made analog.
              ; Set TRISD.
```

This is needed because the 16F877A starts up and resets to analog mode. In analog mode, all of PORTE and PORTA (except pin A.4) are in analog mode. We need the three PORTE pins to be in digital mode so we can control the LCD. The preceding instruction does this. It also makes PORTA digital, but that is not necessary for the LCD.

Go over your code to make sure that all the preceding conditions are met *verbatim* with all upper case and lower case letters correctly typed.

Write a short program to check the operation of the LCD hardware as comprehensively as you think is necessary. The following program can be used as a quickie starter.

Program 18.1 **A short, rudimentary program for testing the LCD**

```
CLEAR                       ; clear variables
DEFINE OSC 4                ; define osc
DEFINE LCD_DREG PORTD       ; define LCD connections
DEFINE LCD_DBIT 4           ;
DEFINE LCD_RSREG PORTE      ;
DEFINE LCD_RSBIT 0          ;
DEFINE LCD_EREG PORTE       ;
DEFINE LCD_EBIT 1           ;
LOW PORTE.2                 ; pull write bit low
                            ;
TRISD = %00000000           ; set all PORTD lines to outputs
TRISE = %00000000           ;
ADCON1 = %00000111          ; don't forget to set ADCON
                            ;
PAUSE 500                   ; pause 0.500 seconds for LCD startup
                            ;
LCDOUT $FE, 1               ; clear LCD, go to first line, first
                            ; position
LCDOUT "Now is the time for" ; print
LCDOUT $FE, $C0             ; go to second line
LCDOUT "a cup of pea soup!"  ; print
END                          ; end program properly
```

A similar but more comprehensive program should be in your utility files to allow you to check the proper hardware and software operation of your LCD whenever you think it is necessary to do so. Your program should check every character and every command in the LCD's vocabulary if you want to perform a really comprehensive check.

Integer math is the source of a lot of problems for those who are unaware of the havoc that integer math calculations can visit on the software you are trying to debug. A certain amount of expertise with integer math is a must if you are going to create mathematical routines within your software. If the routines are amenable to it, you should write a program around the routine to test every possibility that the routine might encounter and thus debug it the hard way, even if it means your computer has to run the routine all night to get through all the commutations. Oftentimes all that is necessary is to run the routines that would be called at the boundary conditions or under the critical conditions to make sure that it is robust.

Here is a routine that has a bug in it. See if you can find it. The problem is designed to make you aware of 8-bit math problems.

We are trying to maintain a value of *X*=127, but we need to do it in small steps. An external condition is changing *X* to anything from 1 to 255 and we cannot exceed these parameters because of the fact that *X* is a variable that has been defined as a BYTE in the program.

```
MODIFYX:                        ;
  IF X<127 THEN X = X+5         ;
  IF X>127 THEN X = X-4         ;
  IF X<1   THEN X = 1           ;
  IF X>255 THEN X = 255         ;
RETURN                          ;
```

Twos compliment convention is not supported in integer math, which is used by the PIC compiler; nor is the implementation of the minus (–) sign.

Configuring the 16F877A and Related Notes

The LAB-X1 is set to 4 MHz by default. This is set by the ABC jumpers on the board. If A is set at 2-3 then it is 4 MHz. Since the compiler writes the default oscillator configuration to XT every time, 4 MHz causes the fewest problems for new users. This means the compiler always uses XT unless you inhibit it. You should inhibit this only if you are not running at 4 MHz. If you are running your own circuit, you will in all probability not be using a frequency dividing network (like the one on the LAB-X1 uses) so your frequency will be the frequency of the crystal that you use. The PIC adjusts the power used by the oscillator to minimize power use (critical for battery powered devices). At the XT setting the PIC puts out less power because at 4 MHz less power is needed. If you are using a 20 MHz crystal, you need to use the HS setting for the oscillator. At the HS setting the PIC provides more power for the oscillator. Sometimes the system will run a 20 MHz crystal at the XT setting but the operation can be marginal.

When working on projects faster then 4 MHz and where power consumption isn't critical, use a 20 MHz crystal. Uncheck the Update Configuration From File option in the programmer after changing to HS. This prevents the setting from reverting to XT every time a new hex file loads (as mentioned above).

The compiler always disables low-voltage programming by default. If you check the option Update Configuration From File in the programming software, this setting will make it to the chip. If you are setting configuration only in the programmer software, simply select Disable for Low-Voltage Programming.

The configuration setting controls how the oscillator driver works on the PIC. It's mainly a question of power consumption, but there also seems to be some filtering involved. If you're using a 4 MHz crystal/resonator, the HS setting will work, but it will consume a bit more power than the XT setting. The XT setting won't drive the faster crystals reliably, presumably because it doesn't have the power.

Here is a situation for which there is no explanation as of now. The LAB-X1 uses an external clock chip, which is essentially the same as a TTL oscillator source. When you use a 16F877 (not 16F877A) set to XT, it will work fine at 20 MHz. If you replace the PIC with a 16F877A, however, it won't run at all. It requires the HS setting for 20 MHz, even though the oscillator source is external and the chip doesn't have to drive a crystal.

If pin B.3 is acting weird and the program randomly stops, starts, or resets, it suggests that you have enabled Low-Voltage Programming in the configuration screen of the programmer. You have to check this whenever you change the PIC you are using because this can change unexpectedly (I could not predict when it would happen).

If the compiler will not erase the chip and gives a "chip not erased" error, it could be that the chip you have selected in the software and the chip you are programming do not match!

Questions and Answers

A few unusual conditions that can arise when we are setting the A and E ports to digital and analog functions with the ADCON1 command are addressed in a question and answer format to address what can happen under special circumstances.

QUESTION

If ADCON1=7, does that mean that the TRIS registers have to be set for ports A and E to make them outputs? What is the effect of following ADCON1=7 with TRISA=0 and PORTA=0 on the PORTA A to D designations? This is important because the compiler does not clear the LCD on startup and can make it hard to see the effect of these commands in complicated circumstances.

ANSWER

The answer depends on what you are trying to accomplish. The TRIS registers are set to $FF on a reset, making all the pins inputs. If you need a pin on PORTA to be an output, you have to clear the corresponding TRIS bit. Commands LCDOUT sets the TRIS values automatically.

```
ADCON1 = 7      ; configures pins for digital operation
TRISA = 0       ; configures pins as outputs (PORTA value
                ; placed on pins)
PORTA = 0       ; changes PORTA value (drives all pins low)
```

If you want to ensure that PORTA has a specific value when it becomes active as an output, set the TRISA value last:

```
ADCON1 = 7      ; configures pins for digital operation
PORTA = 0       ; changes PORTA value (pins still inputs -
                ; floating)
TRISA = 0       ; configures pins as outputs (PORTA value
                ; placed on pins)
```

The reset values of the registers are listed in the data sheet. They differ based on the type of reset, the port, and the PIC you are using. Some bits are set to 1 or 0, while others are specified as being "unknown" or "unchanged."

QUESTION

What if ADCON1 does not set all the A and E ports to digital? If a PORTA pin is still analog and we try to set it high with TRISA, what happens?

ANSWER

This also depends on the PIC and the pin in question, but the following answer will serve in the majority of cases.

The output circuitry is not affected by the pin being configured as analog mode. Therefore you can write to the port as a whole and it will work as expected:

```
PORTA = $03
TRISA = %00000000  ; RA0 and RA1 go high, even if analog
```

The problems occur when reading the pins. A digital read of the port always returns zeros for pins configured as analog. Because of the read-modify-write phenomenon. Successive writes to different pins on a single port can produce unexpected results:

```
TRISA = %00000000     ; set pins as outputs
PORTA.0 = 1           ; RA0 goes high
PORTA.1 = 1           ; RA1 goes high, but RA0 returns low
                      ; unexpectedly
```

The behavior depends on the method used to send output. Avoid making changes right after one another to fix this most of the time! This is a quirk in the PIC.

Settings

The following applies to the Configuration and Option menus in the programmer software:

Oscillator	XT (for 4 MHz crystal)
Watchdog	Enabled
Power Up	Enabled (not critical)
Brownout	Enabled (not critical)
Low-Voltage Programming	Must be disabled
Flash Program Write	Enabled (critical only if using boot-loader)
No Code or Data Protection	(disable)

When we want the run the LAB-X1 at 20 MHz, its fastest speed, the following conditions have to be set in the programmer options menus.

- Set the oscillator to HS for speeds faster than 4 MHz.
- Using a 20 MHz crystal is fine for most PICs. Check the data sheet for each unit.
- Be sure to select the correct device number in the Device ID box in the programmer.

These programmer option selections are reflected in Table 18.1.

SIMPLE CHECKS

Here are some simple checks that you should apply if the PIC still does not oscillate:

- Check that the power is on to the project.
- Check that the power is on to the programmer.
- Check that the programmer is plugged into your board.
- Check that there is a device in the main PIC socket.
- Check that the device is in there and is orientated properly for pin 1.
- Check that the ZIP socket is locked in its position.
- Check that the correct device has been selected.
- Check that the program has an END statement.
- Check that the addresses for all the DEFINEs for the project are correct.
- Check for spelling errors. (The compiler does not catch everything!)
- Check that everything is in capital letters where required (see PBP manual for details).
- Check that all sockets that should be empty are actually empty.
- Run a test program suited to and designed for your specific project.

TABLE 18.1 PROGRAMMER OPTIONS SELECTIONS

Configuration		
	Oscillator	For 4 MHz use XT
		For 20 MHz use HS
	Code protection	Disable
	Watchdog Timer post-scaler	Disabled
	Brown out reset	Disabled
Enable	Watchdog timer enable	
	Power Up Timer enable	Disabled
Enable	Brown out reset enable	
	Low-voltage programmer enable	Disabled
Enable	Flash program write enable	
	External data bus width	Ignore
	Mode	Ignore
	Memory size	Ignore
Options		
Enable	Program/verify code	
Enable	Program/verify configuration	
Enable	Program/verify data	
	Program/verify ID	Ignore
	Program/verify oscillator calibration	Ignore
	Program serial number	Ignore
DISABLE Update Configuration	If not then 4 MHz	
	Reread file before programming	Ignore
Enable	Erase before programming	
Enable	Verify after programming	
	18Fxxx file data address x2	Ignore
Enable	Disable completion messages	
	Skip blank check	Ignore

- Check with an oscilloscope that the system is actually oscillating at the OSC pins.
- In most programs, some other pins should also be going high and low regularly. Check that this is happening.
- Recheck the wiring.
- Check for open wires and cold solder joints.
- Check for short circuits.
- Recheck the program to make sure that it makes sense.
- Run a program on a PIC that you know works. You may have destroyed the PIC. It is also possible to destroy the internal wiring on only some pins on a PIC!

PROGRAMMER RELATED ERROR MESSAGES

When there is a problem with the program that the compiler can decipher, the messages returned by the compiler can still be somewhat cryptic. Here are some explanations that may clarify some the meanings of these messages for beginners

- If the programming connector is not connected, you will get a code check error.
- If the programmer has no power, you will get a communication error.
- If the wrong device is selected, you will get a blanking error.
- Occasionally, even if everything is okay you will get a code check error. Just reprogram the chip.

The following is a list of things I have noticed and have not yet figured out. If you know the answer, send me an e-mail so we can all share the information.

- Sometimes for some reason INTCON.0 can be set at 1 on startup. If you clear it, it does not reset itself.
- On startup OPTION_REG is %11111111, but if you set it to %01111111 the system hangs up if PORTB is not set appropriately to reflect the need for the Option Register setting. That is, if the PORTB pins are to be pulled up, some of them have to be programmed as inputs or the PIC can hang up!
- B7 and B6 (and sometimes B7 alone) are pulled low by the programming cable of the parallel programmer under certain conditions, which can inhibit the use of these two pins by the software while the programmer is connected to the board.
- The Global Interrupt Enable bit is cleared whenever an interrupt is set and reset automatically when the interrupt bit is cleared by the program. This means no new interrupt can be set till you clear the last interrupt. Keep in mind that Timer0, the free running timer, can complicate this if it is turned on. The easiest way to handle this it to set only one interrupt at a time and follow through till it has been cleared before setting another interrupt, maybe even using another timer or external device.
- There can be more than one ON INTERRUPT GOTO call within a program so each interrupt can be handled individually. It is best to stick with one interrupt till you get good at working with the PICs.

When using the LCD the clearing routine at the beginning of the program needs to be something like the following:

```
PAUSE 500       ; pause for LCD start up
LCDOUT $FE, 1, "Clearing the LCD"   ; clear the display and
                                    ; show message.
PAUSE 200       ; this is useful for seeing a reset response
LCDOUT $FE, 1  ; clear again to make sure you are starting
                ; with nothing in the LCD screen.
```

This is necessary because neither the compiler nor the CLEAR command clears the LCD display on reset, and whatever happens to be in the display from the last program will stay in the display and mislead you. Since this can be confusing at the beginning of a program, it is your responsibility to clear this up.

SETTING THE PORTS

All ports that will be used should be set for port configuration, and all pins that will not be used should be made inputs. Setting them as inputs minimizes the possibility that an improperly set port pin will turn something on or off by mistake. It would not be out of line to actually set all other unused ports to inputs, though all ports are set to inputs on startup. To know the exact official status of each port and register on startup and reset, see the data sheet. Some pins and bits may come up as undetermined.

If you don't set the ports, what was left over from the last time the PIC was programmed can show up in a later program. The compiler does not perform a perfect reset on start up of a new compilation. It does not clear the LCD, anything written to EEPROM, or the LCD memory.

19

CONCLUSION

After all is said and done, running motors with a PIC is a matter of putting together a series of components and segments of programs, each of which provides a specific function—not unlike what we have done in almost all the projects in this tutorial.

In this tutorial, we have covered the most basic techniques for doing this. Other, more sophisticated techniques should not be any harder to investigate and assemble. Writing a short program to investigate what needs to be done for any part of your project should not be difficult with the expertise you now have, although incorporating the code into a larger program can get complicated if timer constraints get in the way. Usually, the most difficult task will be getting the programs to run fast enough to get the job done in the time available. However, in this tutorial we ran all the programs at 4 MHz. You will find that considerably more can be done at 20 MHz. Most of the PICs can be run at 20 MHz, and some can be run at 40 MHz.

Repeated calculations and comparisons are time consuming, as is writing to the LCD. Iterations can take up a lot of time and should be avoided. Avoid or at least reduce the number of times that a calculation is performed and the LCD is written to. If a calculation can be done up front, do it and store the result. Avoid doing the same calculation over and over again in a loop.

It is well worth your while to learn how to really use the timers and counters. They are the key to getting a lot of things done right, *fast*, and with the proper timing. Build your programs up a line of code at a time and be sure that you understand exactly what each line of code does.

The 16F877A has 8K of memory. Most of the programs we wrote were in the 400 to 800 word range, so considerably more sophisticated programs can be written without adding any memory. On the other hand, adding one wire memory is neither hard to do nor expensive.

Only a few instructions are used in the projects we undertook. I did this to keep the emphasis on the development of the projects as opposed to learning what wonderful tricks can be done with the language and how powerful the language is. Expanding the number of instructions you are comfortable with will make your projects more powerful. The first half of the book will help you in this direction.

Circuit diagrams are provided for all the projects to help get you comfortable with designing your own projects. As you can see, this is not overly difficult. All the drawings I made are on the web site in AutoCAD format, and you can cut and paste from them to speed up your work.

Oftentimes it might be necessary to use more than one interrupt and have more than one timer or counter in operation. This can get complicated, and I gave no hint on how to proceed when this is the case. At 250-plus pages, this tutorial is already getting too long. Those techniques will have to wait for the next, more advanced, tutorial!

Part III

APPENDIXES

The Appendixes

A

SETTING UP COMPILER FOR ONE KEYSTROKE OPERATION

It is possible to set up the microEngineering Labs programmers for one keystroke programming so that one keystroke (F10) or a mouse click on the Compile and Program icon will do the following:

1. Open the programmer software.
2. Compile the program.
3. Check the program for errors.
4. Send the program you have in you PC to the PIC in the programmer.
5. Shut down and close the programmer window.
6. Run the program in the PIC.

The following instructions outline the process:

1. Open the MicroCode Studio Editor.
2. Pull down the View menu and select PicBasic Options.
3. Under the programmer bar select the programmer you are using as the default programmer.
   Click Edit and enter the name of the programmer file (meprog.exe). Click Next.
4. Let the program find the file automatically. If it cannot, find the file for it with the browse utility. Click Next.
5. In the filename parameters, at the end of the line add a space, -p, another space and -x. The entry now ends in filename -p -x. Do not omit the two spaces in the above before each - sign.
6. Set up to omit completion messages.

Having done this, whenever you want to transfer the program you are editing to the PIC, just press F10 or click the Compile and Program icon.

B

ABBREVIATIONS USED IN THE BOOK AND IN THE DATA SHEETS

The following is an abridged dictionary for the uninitiated.

A to D	Analog to digital.
ADC_BITS	Analog-to-digital converter. Sets the number of bits that the analog-to-digital conversion uses, usually 8 or 10.
ADC_CLOCK	Analog-to-digital converter. Defines where the clock or the process will be read from.
ADC_SAMPLEUS	Analog-to-digital converter. Defines the sample rate in micro seconds (µ is for mu, micro).
ADCIN	Analog-to-digital converter. Instruction to read the analog digital channel input line selection, the identification number that follows tells which line to read.
BASIC	An easy to learn language for programming computers.
BIT	A 1-bit variable. Can hold a number up to 2; that is, a 0 or a 1.
BOR	Brown-out Reset.
BYTE	An 8-bit variable. Can hold a number up to 255.
CLKIN	Clock Input line.
CMOS	Complementary Metal Oxide Semiconductor.
CPU	Central processing unit.
CS	Chip select.
DC	Direct current.
EEPROM	Electrically Erasable Programmable Read Only Memory.
EPIC	Programmers made by microEngineering Labs.

EPROM	Erasable Programmable Read Only Memory.
FLASH	Memory that can be programmed electrically in a flash.
I2C	I two C, a type of serial memory.
IC	Integrated circuit.
ICSP	In-circuit serial programming.
IR	Infrared.
kHz	Kilohertz, 1000 cycles per second.
LAB-X1	Experimental board's name.
LCD	Liquid crystal display.
LCD_DBIT	Liquid crystal display data bit.
LCD_DREG	Liquid crystal display data register.
LCD_EBIT	Liquid crystal display Enable bit.
LCD_EREG	Liquid crystal display Enable register.
LCD_RSBIT	Liquid crystal display Register select bit.
LCD_RSREG	Liquid crystal display Register select register.
LCDOUT	Liquid crystal display Output. Sends the information to the liquid crystal.
LED	Light emitting diode.
mA	Milliamps: 0.001 amps, one thousandth.
μA	Microamps: 0.000001 amps, one millionth.
MCLR	Master clear linen on MCU. Resets the chip on startup.
MCU	Microcontroller unit; the PIC 16F877A is a microcontroller.
MHz	Megahertz: 1,000,000 cycles per second.
Nibble	Half a byte; a 4-bit variable. Can hold a number up to decimal 16.
Ns	Nanosecond: 0.000,000,001 seconds, 10 to –9th.
PCB	Printed circuit board.
PIC	Peripheral interface controller, the original name for microcontrollers.
POR	Power-on Reset.
PRO	Professional.
PROM	Programmable Read Only Memory.
PSP	Parallel Slave Port.
PWM	Pulse width modulation.
PWRT	Power up on Reset Timer.
RAM	Random Access Memory.

RC	Resistor-capacitor, usually used with oscillators.
RISC	Reduced Instruction Set Computer.
RS232	A communications standard for short haul communications.
RS485	A communications standard for longer haul communications.
SPI	A 1-wire communications standard.
SSP	Synchronous serial port.
SST	Oscillator startup timer.
TOCKI	Timer Zero Clock 1.
TRISA	Tri State Register "A."
VAR	A declared variable. All variables used must be declared up front before use.
Vcc	Power for the devices; that is, for the motor you are running.
Vdd	Logic power voltage level for chips as regards supply power.
Vss	Logic ground for chips as regards supply power.
WDT	Watchdog Timer.
WORD	A 16-bit variable. Can hold a number up to 65536.
ZIF	Zero Insertion Force.

C

THE BOOK SUPPORT WEB SITE

A brief description of all files on the book support web site (www.encodergeek.com) and their intended uses:

Text only listing or all microEngineering Labs programs All the programs in one file. Allows you to search for any command you are interested in to see how it was used in a program by microEngineering Labs.

Text only listing of all the programs in the book by chapter All the programs in the book. Allows you to cut and paste them into your work as needed.

Color photographs of various items Reference material.

AutoCAD files Schematic file with all the circuit diagrams in it. AutoCAD file format. For use in creating your own diagrams and designs.

Book reviews Short reviews of books related to PIC microprocessors. Helps you decide on certain very basic-level book purchases.

D

SOURCES OF MATERIALS

Here are the sources of the materials I used, along with some other useful information regarding the projects that were undertaken in the book.

LAB-X1 Board

microEngineering Labs,
Box 60039
Colorado Springs, CO 80960
(719) 520-5323

Motors

Stepper motor $2.49
Alltronics.com
Item: 12V 12.8 Ohm - Stepper Motor - 98M001
4-wire 12 VDC @ 0.9A, 100s/rev, 3.6°, with 2" x 0.25" lead screw. NEMA-17 size.

Servomotors with Encoders

Available as four models.
Encodergeek.com
705 W. Kirby
Champaign, IL 61820
(217) 359-6751

R/C Servos

HS 311 Standard servo $8.99
Tower Hobbies
towerhobbies.com
1602 Interstate Drive
Champaign, IL 61822
(800) 637-4989

Amplifiers

Xavien
Two different amplifiers available, one axis and two axis.
xavien.com
P. O. Box 7433
Goodyear, AZ 85338

Solarbotics

Amp is "the L298 motor driver kit."
Solarbotics.com
201-35th Avenue NE
Calgary, AB T2E 2K5
Canada
(866) 276-3687

Controller from Encodergeek.com

As a part of writing this book, I designed, assembled, and programmed a comprehensive controller suitable for running the DC motor described in this book, to prove to myself that everything works as described.

This Servomotor controller can be used with motors with or without encoders. Incorporates the features discussed in the book and is ready to be played with by you. It is designed specifically for experimentation, It is completely programmable with microEngineering Labs programmers, and can use PBP language and compiler or use assembly language, "C," and so on.

Features include:

- Up to 3 amps at 55 V 18200 chip H bridge
- Processor, uses PIC 18F4331 or 16F877A, can use others
- Two onboard pots
- Real-time potentiometer control
- R/C control from radio
- PWM control
- Encoder interface matches motors in this book
- Reversing switch
- Microswitch for over travel
- Microswitch for under travel
- Two lines of 16-character display
- Onboard power regulation
- Separate inputs for logic and motor power
- Programming switches (16 choices, 4 bits)
- Power switch
- 20 MHz operation

- Two programmable LEDs for indication
- Power LED
- Main switch
- 16 modes of operation
- Programming connector
- Matches microEngineering Labs standard
- Source code provided

This controller is suitable for all your experiments for running DC motors with and without encoders. The two onboard potentiometers allow you to change to variables without having to reload a program. This can be a tremendous time saver as you experiment with your software. (However, it does not run stepper motors.)

The controller is powered from two unregulated 12 VDC wall transformers. One provides logic power and the other provides power for the motor. Both need to be 2.1 mm units with positive center connectors. These transformers are what most electronic devices on the market come with. You should already have a couple in your junk drawer.

Cost: \$85.00 postpaid in the USA; \$99.00 postpaid in the USA with large encoded motor and 12" encoder cable(prices circa 2009).

The motor is available from Encodergeek.com, my web site.

The controller is illustrated in Figure D.1.

The PORT designations and schematics follow.

Figure D.1 Experimentation board: motor controller for DC motors, with or without an encoder

PORT DESIGNATIONS

The board is wired with the following address designations for the PIC 18F4331:

PWM input	PORTA.0
Left potentiometer	PORTA.1
Right potentiometer	PORTA.2
Encoder A	PORTA.3 Set pot to middle when using LAB-X1
Encoder B	PORTA.4
~~Not used~~	~~PORTA.5~~
microEngineering Labs programmer	PORTB.7
microEngineering Labs programmer	PORTB.6
Programming switch 1	PORTB.5
Programming switch 2	PORTB.4
Micro Switch dir fwd	PORTB.3
Micro Switch dir back	PORTB.2
Micro Switch REVERSER	PORTB.1
Pulse from R/C signal	PORTB.0
Motor brake	PORTC.0 = 0 to turn brake off
Motor PWM	PORTC.1
~~Not used~~	~~PORTC.2~~
Motor direction	PORTC.3
Selection switch 3	PORTC.4
Selection switch 2	PORTC.5
Selection switch 1	PORTC.6
Selection switch 0	PORTC.7
~~Not used~~	~~PORTD.0~~
~~Not used~~	~~PORTD.1~~
Green LED	PORTD.2
Red LED	PORTD.3
LCD data	PORTD.4
LCD data	PORTD.5
LCD data	PORTD.6
LCD data	PORTD.7

SETTING UP FOR THE 18F4331

When programming the 18F4331, the programmer options can be set to the selections listed in the following table. These may have to be changed is your program includes features that were not used in the programs in this book.

Oscillator	HS
Int. external switchover	Disabled
Fail safe clock monitor	Disabled
Power up timer	Disabled
Brown out reset	Disabled
Brown out reset voltage	2.0 V
Watchdog Timer	Disabled
Watchdog Timer post-scaler	1:32768
Watchdog Timer window	Enabled
PWM pins	Disabled on reset
Low side transistor polarity	Active High
High side transistor polarity	Active High
Special event reset	Disabled
FLTA input multiplexed with	RC1
SSP I/O multiplexed with	RC4, 5, 7
PWM 4 multiplexed with	RB5
External clock multiplexed with	RC3
MCLR pin function	Reset
Stack underflow overflow reset	Enabled
Low voltage programming	Enabled
Boot block	Not protected
Codes and all the rest	Not protected

The wiring schematic for the controller board is provided in Figure D.2

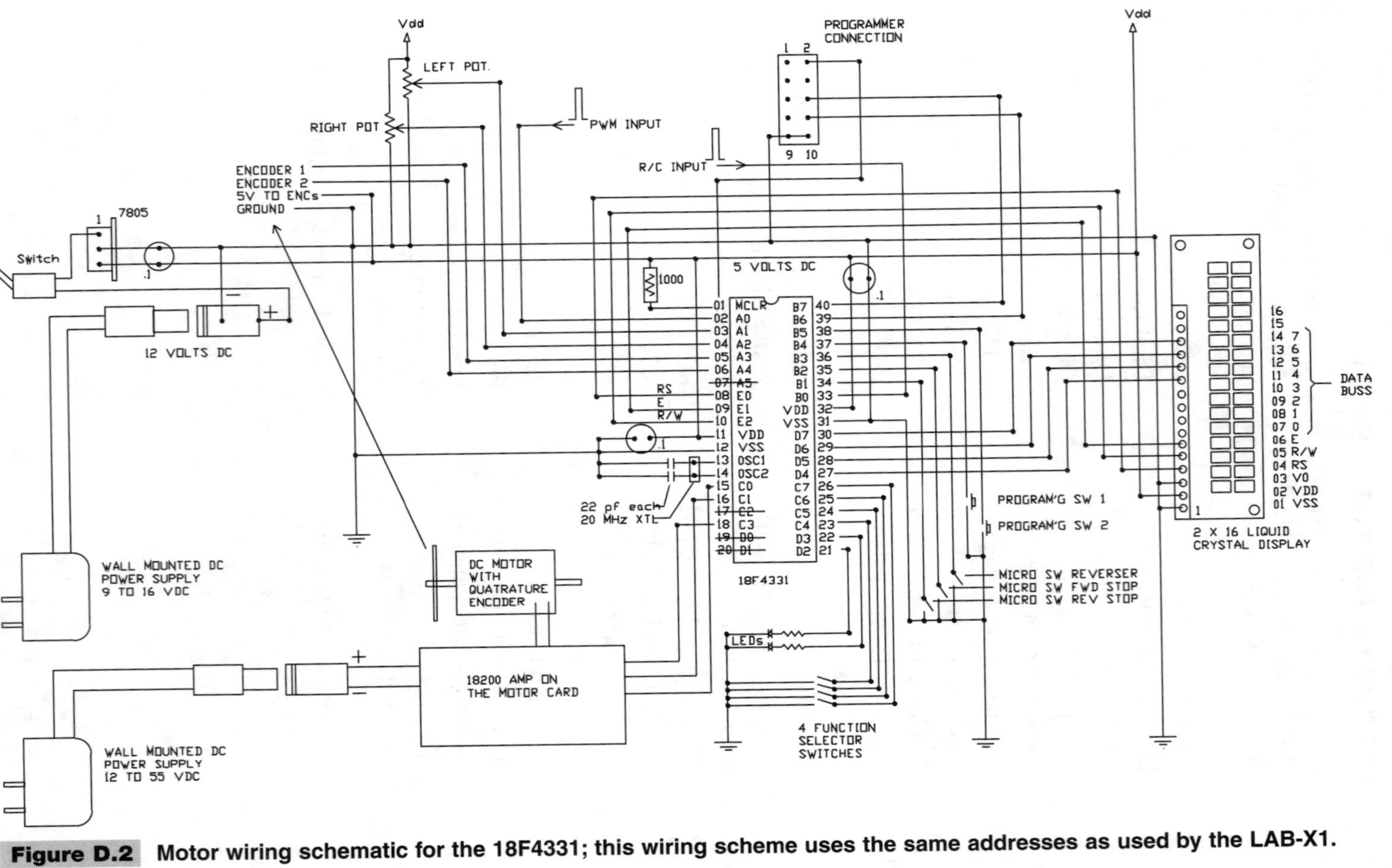

Figure D.2 Motor wiring schematic for the 18F4331; this wiring scheme uses the same addresses as used by the LAB-X1.

MOTOR CONTROL LANGUAGE: SOME MINIMAL IDEAS, GUIDANCE, AND NOTES

Further down the road, you might want to write a language to control the motor you are using. Here some ideas on how you might go about assigning the alphabet to motor functions. Though in no way a comprehensive approach, it might give you some ideas as to how to proceed. These are the kinds of commands that are needed to control a motor.

Language Commands

Cmd.	Meaning	Bytes	Value range
A	acceleration and setting	2	0 to +,– 32000
B	brake immediate	0	0 or 1
C	coordinated move set	0	0 or 1
D			
E			
F	factory defaults set	0	0 or 1
G			
H			
I			
J			

K	move status (doing or done)	0	0 or 1
L	loops	2	0 to +,– 32000
M	motor identification	1	8 bits in a byte
N			
O			
P	power value setting	2	0 to +,– 32000
Q	power mode set	0	0 or 1
R	run	0	0 or 1
S	stop	0	0 or 1
T	trapezoidal mode set	0	0 or 1
U			
V	velocity mode set	2	0 to +,– 32000
W			
X	erase everything to 0	0	0 or 1
Y			
Z	sleep set	2	0 to +,– 32000

Interrogation

When you write a comprehensive control language, you need a way to interrogate the controller to know what is going on in the motor/controller system from time to time. The information that you need will have been stored in certain memory locations in the controller. These registers represent the state of the machine at any one time. The following information is designed to start you thinking of what might be implemented in your control language.

All registers in the system should be readable in real time. When asked for, the information is sent to the controlling computer where it is interpreted and the next command is issued. The registers needed in a typical system contain information like:

- Current position
- Target position
- Next position
- Motor gain
- Switch positions

What you store in the various registers will depend on what is needed to control the motor and to allow the information to be requested by your PC to be sent to it as rapidly as possible.

The purpose of writing a language is to allow you to use a computer to control the motor. What the language does depends on what you want the language to do for the specific task that you have in mind. Since the language resides in the computer, it can be as large as you need it to be. Since the language interpretation is done in the controller, it has to be pretty compact.

As was mentioned earlier, some look-ahead capability makes for smoother move transitions. It is important that the motor not stop between moves as the speed is changed from instruction to instruction. This is necessary for almost all applications.

If you decide not to use the RS 274D language, and it is necessary to follow this language if you want some compatibility with an existing standard, it might be possible to design a more efficient and faster system.

An Industrial Language Used by CNC Machines

There is a standard language that is used to control CNC machines of all kinds. This language does not have a formal name, but it does have an official designation. It is the RS-274D standard by EIA and is often referred to as the G and M codes language. Most manufacturers implement a dialect of this language.

RS-274D is the standard for numerically controlled machines developed by the Electronic Industry Association in the early 1960s. The RS-274D revision was approved in February 1980.

There are a number of historical sidelights to this standard, many having to do with the original use of punched paper tape as the only data interchange medium. The 64-character EIA-244 paper tape standard is now (thankfully) obsolete, and ASCII character bit patterns are now the standard representation. Others are methods for searching for specific lines (program blocks) on the tape, rewinding the tape, and so on.

The basic unit of the program is the "block," which is seen in printed form as a line of text. These lines usually start with a number, such as N0001 X123.

Each block can contain one or more "words," which consist of a letter describing a setting to be made, or a function to be performed, followed by a numeric field supplying a value to that function. An example would be X10.001, which by itself indicates the X axis should move to a position of 10.001 user units, which would normally be inches or mm. Various words can be combined to specify multiaxis moves or perform special functions.

The common axes are normally named the following:

- **A** Angular axis around X axis
- **B** Angular axis around Y axis
- **C** Angular axis around Z axis
- **U** Secondary axis parallel to X
- **V** Secondary axis parallel to Y

- **W** Secondary axis parallel to Z
- **X** Primary linear axis
- **Y** Primary linear axis
- **Z** Primary linear axis

Control words are the following:

- **F** Feed rate
- **G** Preparatory functions
- **M** Miscellaneous function
- **S** Spindle speed
- **T** Tool function

The preparatory (G) functions are as follows:

- **G00** Positioning
- **G01** Linear interpolation
- **G02** Circular (clockwise) interpolation
- **G03** Circular (counterclockwise) interpolation
- **G04** Dwell (not modal)
- **G17** X-Y plane
- **G18** Z-X plane
- **G19** Y-Z plane
- **G33** Thread cutting, constant lead
- **G34** Thread cutting, increasing lead
- **G35** Thread cutting, decreasing lead
- **G40** Cancel cutter compensation
- **G41** Cutter compensation, tool left of path
- **G42** Cutter compensation, tool right of path

How to use cutter diameter compensation

- **G43** Tool length offset

How to use tool length offset

- **G49** Cancel tool length offset
- **G70** Inch programming
- **G71** Metric programming

The miscellaneous (M) functions are as follows:

- **M00** Program Stop
- **M01** Optional program stop
- **M02** End of program

- **M03** Spindle CW
- **M04** Spindle CCW
- **M05** Spindle stop
- **M06** Tool change
- **M07** Flood coolant on
- **M08** Mist coolant on
- **M09** Coolant off

The preceding information is from the Internet. Search on "RS-274D" for more detailed information.

INDEX

My Table of Contents

Keep track of your frequently accessed pages here.